Nonlinear Waves

Theory, computer simulation, experiment

Series on wave phenomena in the physical sciences

Series Editor
Sanichiro Yoshida
Southeastern Louisiana University

About the series

The aim of this series is to discuss the science of various waves. It consists of several books, each covering a specific subject known as a wave phenomenon. Each book is designed to be self-contained so that the reader can understand the gist of the subject. From this viewpoint, the reader can read any book as a stand-alone article. However, it is beneficial to read multiple books as it would provide the reader with the opportunity to view the same aspect of wave dynamics from different angles.

The targeted readership is graduate students of the field and engineers whose background is similar but different from the subject. Throughout the series, it is intended to help students and engineers deepen their fundamental understanding of the subject of wave dynamics. An emphasis is laid on grasping the big picture of each subject without dealing with detailed formalism, and yet understanding the practical aspects of the subject. To this end, mathematical formulations are simplified as much as possible and applications to cutting edge research are included. The reader is encouraged to read books cited in each book for further details of the subject.

Other titles in this series

William Parkinson *What's the Matter with Waves? An Introduction to Techniques and Applications of Quantum Mechanics*

Sanichiro Yoshida *Waves: Fundamentals and Dynamics*

Wayne D Kimura *Electromagnetic Waves and Lasers*

David Feldbaum *Gravitational Waves*

Nonlinear Waves

Theory, computer simulation, experiment

M D Todorov
Technical University of Sofia, Bulgaria

Morgan & Claypool Publishers

ISBN   978-1-64327-047-0 (ebook)
ISBN   978-1-64327-044-9 (print)
ISBN   978-1-64327-045-6 (mobi)

DOI   10.1088/978-1-64327-047-0

Version: 20180801

IOP Concise Physics
ISSN 2053-2571 (online)
ISSN 2054-7307 (print)

A Morgan & Claypool publication as part of IOP Concise Physics
Published by Morgan & Claypool Publishers, 1210 Fifth Avenue, Suite 250, San Rafael, CA, 94901, USA

IOP Publishing, Temple Circus, Temple Way, Bristol BS1 6HG, UK

To my family

Contents

Preface

In this book, we study a few dynamical models of nonlinear solitary waves. In chapter 1, we start with the Boussinesq equation (derived in 1871), which was the first model of surface waves in shallow water that has in mind the balance between the nonlinearity and dispersion keeping the wave shape (wave of transition). These waves behave as particles called solitons by Zabusky and Kruskal in 1965 [1] (collision property) in special cases. The discovery that these localized solutions (with finite energy in an infinite region) can keep their individuality after a collision brought us to the concept of quasi-particles. This concept can be very important when one interprets the dualism wave-particle in physics. In 1D, a plethora of deep mathematical results have been obtained for solitons (see [2, 3]). The success was contingent upon the existence of an analytical solution of the respective nonlinear dispersive equation. The prevalent part of the known analytical and numerical solutions of the Boussinesq equations relates to a 1D case, while for multidimensional cases almost nothing is known so far. The difficulties originate from the lack of known analytic initial conditions and the nonintegrability in the multidimensional case. An exclusion is the solutions of the Kadomtsev–Petviashvili equation (KPE) that can be considered as a reduction of the Boussinesq equation and has a limited validity. The KPE has fourth derivatives only in one of the spatial directions, while in the other direction, the highest-order is second. Interesting analytical results are obtained for its solutions, which are localized in the direction with the fourth-order derivative, and are periodic in the other direction (see, e.g. [4–6] and the literature cited therein).

Apart from the Boussinesq equation, of special interest are the generalized wave equations with nonlinearity and dispersion, like the nonlinear Schrödinger equation and the nonlinear evolution equations like the Korteweg–de Vries equation. Chapter 2 is devoted to one generalization of the scalar nonlinear Schrödinger equation—the dynamical system of coupled nonlinear Schrödinger equations known also as the vector Schrödinger equation. This system possesses very rich phenomenological subjects of investigation. In the general case, it is a very useful spring-board to study the dynamics of the solitary waves as quasiparticles when the integrability is reduced to only three conservation laws. An important advantage is the polarization dynamics of the waves. All of these properties of the vector Schrödinger equation as well as of his particular case—the Manakov system—are studied, described and discussed in detail.

The last chapter, chapter 3, is an extension for the multidimensional case of the scalar nonlinear Schrödinger equation, as well as envelope equations that describe the dynamics of ultrashort light pulses. This chapter relates to the experimental data and comparisons and tries to reveal less studied properties of the matter. The context here is again in soliton-solutions existence and their spatiotemporal stability.

More physical systems are not fully integrable even in the 1D case and the main tool to investigate and understand the pertinent physical mechanisms of the interactions is the numerical one. We maintain the attitude that the real nonlinear nature of the dynamical systems under consideration is displayed when they lose their integrability. In the author's opinion, they then justify the confidence to be called and considered as real nonlinear dynamical models. The resolving of such kind of problems requires inseverable combination and complementing (not contra-distinction) of analytical and numerical approaches, methods, and techniques. The nonlinear problems require a nonlinear behavior of research in order to be resolved successfully.

References

[1] Zabusky N J and Kruskal M D 1965 Interaction of "solitons" in a collisionless plasma and the recurrence of initial states *Phys. Rev. Lett.* **15** 240

[2] Ablowitz M J and Segur H 1981 *Solitons and the Inverse Scattering Transform* (Philadelphia, PA: SIAM)

[3] Newell A C 1985 *Solitons in Mathematics and Physics* (Philadelphia, PA: SIAM)

[4] Christov C I, Maugin G A and Porubov A 2007 On Boussinesq's paradigm on nonlinear wave propagation *C. R. Mecanique* **335** 521–35

[5] Porubov A V, Maugin G A and Mateev V V 2004 Localization of two-dimenasional non-linear strain waves in a plate *Int. J. Nonlinear Mech.* **39** 1359–70

[6] Porubov A V, Pastrone F and Maugin G A 2004 Selection of two-dimensional nonlinear strain waves in micro-structured media *C. R. Mecanique* **332** 513–8

Acknowledgements

The investigations were conducted in the last 10 years in close collaboration with Professor C I Christov (University of Louisiana at Lafayette, LA, USA), Professor V S Gerdjikov (Institute of Nuclear Research and Nuclear Energy, BAS, Sofia, Bulgaria), Professor I G Koprinkov (Technical University of Sofia, Bulgaria). Part of them was supported by the Bulgarian Science Foundations under two grants—http://oldweb.tu-sofia.bg/fond-NI/klasirani-proekti/FPMI-DDBU02-71/project.htm or http://www.math.bas.bg/nummeth/boussinesq/, and http://www.math.bas.bg/~nummeth/nonlinear/index.htm. We attracted researchers from the Technical University of Sofia, Bulgarian Academy of Sciences, St Kliment Ohridski University of Sofia, University of Nicosia, Cyprus, etc. All of them are acknowledged for the successful investigations. A major part of the results presented here are based on published articles in journals and referred issues of Elsevier, AIMS, AIP, Springer, MATEC, etc. The narrative is a study and generalization of these publications.

Author biography

M D Todorov

Michail Todorov graduated in 1984 and received a PhD degree in 1989 from the St. Kliment Ohridski University of Sofia, Bulgaria. Since 1990, he has been Associate Professor and Full Professor (2012) with the Department of Applied Mathematics and Computer Science by the Technical University of Sofia, Bulgaria. He has worked as a Senior Research Fellow in the Joint Institute for Nuclear Research at Dubna, Russia (2004) and as a Visiting Professor, a Visiting Scholar, and a Visiting Consultant in the University of Texas at Arlington, TX, USA (2008, 2009 and 2011) and Texas A&M University at Commerce, TX, USA (2011), Sabbatical Professor at Southeastern Louisiana University at Hammond, LA, USA (2013) and Embrie-Riddle Aeronautical University, Daytona Beach, FL, USA (2017). Since 2000, he has also been a part-time employed instructor on Computer Science and Technology in the St. Kliment Ohridski University of Sofia, Bulgaria. In 2004–2008, he was a part-time employed instructor on theoretical electrodynamics in the Paisii Khilendarski University of Plovdiv, Bulgaria.

For the last few years, his primary research areas have been mathematical modeling, computational studies, and scientific computing of nonlinear phenomena including soliton interactions, nonlinear electrodynamics, nonlinear optics, mathematical biology and bioengineering, and astrophysics.

Dr Todorov is an editor of eleven peer reviewed books in the Conference Proceedings Series by the American Institute of Physics, Melville, NY, USA, and Guest-Editor in Wave Motion (Elsevier) and Springer Proceedings. He is a coordinator, chair and/or special session organizer, member of Program and Scientific Committees of about 40 conferences on Applied Mathematics, Dynamical Systems and Differential Equations in Bulgaria, Russia, USA, and Taiwan. More information can be found by visiting his homepage at http://2014.eac4amitans.eu/MTodorov/index1.htm.

IOP Concise Physics

Nonlinear Waves
Theory, computer simulation, experiment
M D Todorov

Chapter 1

Two-dimensional Boussinesq equation. Boussinesq paradigm and soliton solutions

The main priorities and difficulties in the study of the Boussinesq equation and in particular in the study of soliton solutions evolve from a variety of 1D versions, as well as from the nonintegrability and the lack of an exact formulation of an initial-value problem of the 2D versions. The investigation is subject to the following items
- Choice of the proper equation;
- Formulation of the Cauchy problem;
- Soliton solutions and their stability. The Boussinesq paradigm;
- Non-integrability and self-consistence;
- Physical and mechanical interpretation and explanation of the results.

1.1 Boussinesq equations. Generalized wave equation

The Boussinesq equation is the first model of surface waves over shallow water that considers the nonlinearity and the dispersion and their interaction as a reason for wave stability, properly termed nowadays as the 'Boussinesq paradigm' (for details, see [28, 32]). Boussinesq [8] proved that the balance between the steepening effect of the nonlinearity and the flattening effect of the dispersion maintains the shape of the wave. This balance bears solitary waves that behave like quasi-particles. He derived the equation and found an analytical solution. We call it the 'original Boussinesq equation' (OBE)

$$u_{tt} = (u - \alpha u^2 + \beta u_{xx})_{xx}.$$

OBE is fully integrable but incorrect in the sense of Hadamard when the dispersion parameter $\beta > 0$. The mere change of the incorrect sign of the fourth derivative in OBE ($\beta < 0$) yields the so-called 'good (proper)' Boussinesq equation (PBE), which is correct in the sense of Hadamard. A different approach to removing the incorrectness is by changing the spatial fourth derivative to a mixed fourth

doi:10.1088/978-1-64327-047-0ch1

derivative, i.e. $\frac{\partial}{\partial t} \approx \frac{\partial}{\partial x}$ (called a 'linear impedance relation'), which results in an equation known nowadays as the regularized long wave equation (RLWE) or Benjamin–Bona–Mahony equation

$$u_{tt} = (u - \alpha u^2 + \beta u_{tt})_{xx}. \tag{1.1}$$

Over the years, OBE was 'improved' in a number of works. Also, the linear impedance relation combined with different kinds of nonlinearity produced innumerable instances of unphysical results. The most popular are the versions that contain a quadratic nonlinearity, which are useful from the paradigmatic point of view. The actual nonlinearity is important because it provides for the Galilean invariance of the model.

Let us emphasize that the above considered versions of the Boussinesq equation admit a sech-like localized solution. For instance, the solution of OBE reads

$$u = -\frac{3}{2}\frac{c^2 - 1}{\alpha} \operatorname{sech}^2\left(\frac{x - ct}{2}\sqrt{\frac{c^2 - 1}{\beta}}\right).$$

Let us also emphasize that the more analytical and numerical results obtained so far concern 1D problems, while in 2D to obtain analytical results is problematic. All of 2D Boussinesq equations are nonintegrable.

As shown in [20], the consistent implementation of the Boussinesq method yields the following generalized wave equation (GWE) for function $f = \phi(x, y, z = 0; t)$:

$$f_{tt} + 2\beta\nabla f \cdot \nabla f_t + \beta f_t \Delta f + \frac{3\beta^2}{2}(\nabla f)^2 \Delta f - \Delta f + \frac{\beta}{6}\Delta^2 f - \frac{\beta}{2}(\Delta f)_{tt} = 0 \tag{1.2}$$

with Hamiltonian density

$$\mathcal{H} = \frac{1}{2}\left[f_t^2 + (\nabla f)^2 - \frac{1}{4}\beta^2(\nabla f)^4 + \frac{1}{6}\beta(\nabla f)^2 + \frac{1}{2}\beta(\nabla f_t)^2\right].$$

The motion in a inviscid shallow layer with free surface is governed by the Laplace equation for the potential ϕ. Keep in mind, the kinematic and dynamic conditions equation (1.2) is a correct energy-conserving form with respect to a new function $f(x, y, t) := \phi(x, y, z = 0, t)$ (see [20] for details).

Equation (1.2) is the most rigorous amplitude equation that can be derived for the surface waves over an inviscid shallow layer, when the length of the wave is considered large in comparison with the depth of the layer. It was derived only in 2001. Besides it, however, a plethora of different inconsistent Boussinesq equations are still vigorously investigated. Let us consider the equation when the velocity potential and the surface elevation do not depend on the coordinate y. Bear in mind that f is the velocity potential on the bottom of the layer where we introduce a vertical velocity $u = f_x$ and an auxiliary function q. Then we obtain a so-called dispersive wave system (DWS), which can be considered as a progenitor of the different 1D Boussinesq equations:

$$u_t + \frac{\alpha}{2}(u^2)_x = q_{xx},$$

$$q_t + \alpha u q_x = u - \beta_2 u_{xx} + \beta_1 u_{tt}.$$

(1.3)

Here, β_1 and β_2 are dispersion coefficients, $\beta_1 = 3\beta_2 = \beta$, and $\alpha = \beta$ is an amplitude parameter.

If we replace the derivative f_t with f_x in the quadratic nonlinear term in (1.2), then we get another implementation of Boussinesq equations called the Boussinesq paradigm equation (BPE)

$$u_{tt} = \left(u + \alpha u^2 + \frac{\beta}{2}u_{tt} - \frac{\beta}{6}u_{xx}\right)_{xx}.$$

(1.4)

Though BPE is a kind of simplification of DWS, it is not invariant with respect of Galilean transformations. The equation is not derived by Boussinesq himself. The 2D Boussinesq paradigm looks like

$$u_{tt} = \Delta(u - \alpha u^2 + \beta_1 u_{tt} - \beta_2 \Delta u)$$

(1.5)

where $u = u(x, y)$ is the surface elevation, Δ is 2D Laplace operator, $\beta_1, \beta_2 > 0$ are dispersion coefficients, and α is an amplitude parameter. We turn to it because almost nothing is known about its nonsteady solutions, including the interactions between soliton solutions (for example, see [1, 44]). No analytic solutions are known to set as exact initial conditions because 2D BPE is not integrable.

1.2 Investigation of the long-time evolution of localized solutions of a dispersive wave system

Preliminary

We consider the long-time evolution of the solution of an energy-consistent system with dispersion and nonlinearity, which is the progenitor of the different Boussinesq equations. Unlike the classical Boussinesq models [6–8], the new one possesses Galilean invariance. As an initial condition, we use a wave system comprised by the superposition of two analytical soliton solutions. We use a strongly dynamical implicit difference scheme with internal iterations, which allows us to follow the evolution of the solution at very long times. We focus on the behavior of traveling localized solutions developing from critical initial data. The main solitary waves appear virtually non-deformed from the interaction, but additional oscillations are excited at the trailing edge of each one of them. We track their evolution for very long times when they tend to adopt a self-similar shape. We test a hypothesis about the dependence on time of the amplitude and the support of Airy-function shaped coherent structures. The investigation elucidates the mechanism of evolution of interacting solitary waves.

Over the years, a plethora of different Boussinesq equations have been derived under the assumption of balance between weak nonlinearity and weak dispersion. They are generalized wave equations which offer the opportunity to investigate the generic features of dispersive wave models, such as head-on collisions of localized structures (solitary waves/quasi-particles) even beyond the framework of the long-wave weakly-nonlinear assumptions (see, for example, [2–5, 50]). Boussinesq equations as a rule possess three standard conservation/balance laws—for mass, energy, and momentum. Recently, a more general form (preserving the Galilean invariance) of the dispersive shallow water equations has been derived [18–21]. It has been shown to possess a solitary-wave solution of sech type, which makes it very useful in a paradigmatic sense for investigation of solitonic (pseudo-particle) behavior of localized solutions. A special staggered conservative difference scheme was constructed in order to meet the case of Galilean invariant systems. A number of cases of soliton interactions were conducted [18, 20], ranging from weakly nonlinear case to a strongly nonlinear case with nonlinear blow-up of the solution in finite time. The model is expected to provide an additional basis for soliton research concerning the long and very long-times tracking of the soliton dynamics, as well as to get more information about the possible self-similar behavior. This is the main deal of the present work.

Problem formulation

In order to find the correct energy-conserving form of equation (1.2), we introduce a new auxiliary function χ and present it as a system

$$
\chi_t = -\beta\nabla \cdot \chi\nabla f - \Delta f + \frac{\beta}{6}\Delta^2 f - \frac{\beta}{2}(\Delta f)_{tt}
$$

$$
f_t = -\frac{\beta}{2}(\nabla f)^2 - \chi.
$$

(1.6)

We call equations (1.6) the 'energy-consistent Boussinesq paradigm.'

Then, we derive the dispersive wave system (DWS) (1.3) substituting for $\chi = -q_x$ in (1.6) [29]. Here, the vertical (dimensionless) velocity $u = f_x$.

The boundary conditions are

$$
u\mid_{x=-L_1, L_2} = 0, \qquad q_x\mid_{x=-L_1, L_2} = 0.
$$

(1.7)

When the interval $[-L_1, L_2]$ is finite, they provide the conservation of the total energy.

We construct the initial condition is a superposition of two solitary waves

$$
u(x, t = 0) = \frac{a\,\text{sgn}(c)}{\dfrac{|c| - 1}{2} + \cosh^2[b(x - X - ct)]}, \qquad a = \frac{c^2 - 1}{\alpha}, \quad b = \sqrt{\frac{c^2 - 1}{2(\beta_1 c^2 - \beta_2)}}
$$

traveling in opposite directions with phase velocities c_l and c_r starting from positions X_l and X_r.

For simplicity, we introduce a new parameter β such that $\beta_1 = \beta/2$, $\beta_2 = \beta/6$, and $\alpha = \beta$. Then, the sech-like solutions exist in two domains—subcritical (subsonic): $0 < c < 1/\sqrt{3}$ and supercritical (supersonic): $c > 1$.

Hamiltonian representation

We define 'mass', M, (pseudo)momentum, P, (pseudo)energy, E:

$$M \stackrel{\text{def}}{=} \int_{-L_1}^{L_2} u\, dx, \quad P \stackrel{\text{def}}{=} \int_{-L_1}^{L_2} u(q_x - \beta q_{xt})\, dx,$$

$$E \stackrel{\text{def}}{=} \frac{1}{2}\int_{-L_1}^{L_2} \left(u^2 + q_x^2 + \frac{\beta}{2}u_t^2 + \frac{\beta}{6}u_x^2 - \frac{\beta}{3}u^3 \right) dx. \tag{1.8}$$

Here, $-L_1$ and L_2 are the left end and the right end of the interval under consideration.

When boundary conditions, equation (1.7), are imposed, the following conservation/balance laws hold

$$\frac{dM}{dt} = 0, \quad \frac{dP}{dt} = \frac{1}{2}\int_{-L_1}^{L_2} \left(\frac{\beta}{2}u_t^2 - \frac{1}{2}u^2 + \frac{\beta}{6}uu_{xx} + q_x^2 \right) dx, \quad \frac{dE}{dt} = 0. \tag{1.9}$$

Numerical method

To solve the Galilean invariant case treated here, we construct a strictly conservative implicit staggered scheme, which is inevitably nonlinear:

$$\frac{u_i^{n+1} - u_i^n}{\Delta\tau} = \frac{q_{i+1}^{n+1/2} - 2q_i^{n+1/2} + q_{i-1}^{n+1/2}}{h^2} - \frac{\beta}{4h}\left[\left(u_{i+1}^{n+1}\right)^2 - \left(u_{i-1}^{n+1}\right)^2 \right],$$

$$\frac{q_i^{n+1/2} - q_i^{n-1/2}}{\Delta\tau} = \frac{u_i^{n+1} + u_i^{n-1}}{2} - \frac{\beta}{2h}\left(q_{i+1}^{n+1/2} - q_{i-1}^{n+1/2} \right)\left(u_i^{n+1} + u_i^{n-1} \right)$$

$$+ \frac{\beta}{2\tau^2}\left(u_i^{n+1} - 2u_i^n + u_i^{n-1} \right) - \frac{\beta}{6h^2}\left(u_{i+1}^{n+1} - 2u_i^{n+1} + u_{i-1}^{n+1} \right).$$

We introduce regular meshes (x_i, τ^n) and $(x_i, \tau^{n+\frac{1}{2}})$ within the interval $[-L_1, L_2]$, $x_i = -L_1 + (i-1)h$, $h = (L_1 + L_2)/m$, $i = 1, \ldots, m$, and $\tau^n = n\Delta\tau$, $\tau^{n+\frac{1}{2}} = \left(n + \frac{1}{2}\right)\Delta\tau$, $n = 0, 1, 2 \ldots$ where m is the total number of grid points. To achieve an implicit approximation of the nonlinear terms, we use the simplest linearization combined with an internal iteration (referred to by the composite superscript k). The difference scheme appears to be robust enough and economical:

$$\frac{u_i^{n+1,\,k} - u_i^n}{\Delta\tau} = \frac{q_{i+1}^{n+1/2,\,k} - 2q_i^{n+1/2,\,k} + q_i^{n+1/2,\,k}}{h^2}$$

$$- \frac{\beta}{8h}\left[\left(u_{i+1}^{n+1,\,k-1}\right)^2 - \left(u_{i-1}^{n+1,\,k-1}\right)^2 + \left(u_{i+1}^n\right)^2 - \left(u_{i-1}^n\right)^2\right]$$

$$\frac{q_i^{n+1/2,\,k} - q_i^{n-1/2}}{\Delta\tau} = -\frac{\beta}{8h}\left(q_{i+1}^{n+1/2,\,k-1} - q_{i-1}^{n+1/2,\,k-1} + q_{i+1}^{n-1/2} - q_{i-1}^{n-1/2}\right)$$

$$\times\left(u_i^{n+1,\,k} - u_i^{n-1}\right) \qquad\qquad (1.10)$$

$$- \frac{\beta}{12h^2}\left[\left(u_{i+1}^{n+1,\,k} - 2u_i^{n+1,\,k} + u_{i-1}^{n+1,\,k}\right)\right.$$

$$\left. + \left(u_{i+1}^{n-1} - 2u_i^{n-1} + u_{i-1}^{n-1}\right)\right]$$

$$+ \frac{\beta}{2}\frac{u_i^{n+1,\,k} - 2u_i^k + u_i^{n-1}}{\Delta\tau^2} + \frac{u_i^{n+1,\,k} + u_i^{n-1}}{2}.$$

The internal iterations start from the functions obtained at the previous time stage $u^{n+1,0} = u_i^n$ and $q^{n+1/2,0} = q_i^{n-1/2}$ and are terminated at certain $k = K$ when

$$\max_i\left|u_i^{n+1,\,K} - u_i^{n+1,\,K-1}\right| \leqslant 10^{-13}\max_i\left|u_i^{n+1,\,K}\right|.$$

The value 10^{-13} is selected to be large enough in comparison with the round-off error. In general, the number of iterations K (in our calculations, we keep them around six to eight) depends on the size of time increment $\Delta\tau$. The linearized scheme has an inextricably coupled seven-diagonal banded matrix. We prove that the above approximation secures the conservation of energy on a difference level for arbitrary potential $U(u)$ (see, for example, [26]), i.e. the difference approximations of the mass and the energy

$$M^{n+1} = h\sum_{i=2}^{N-1} u_i^{n+1},$$

$$E^{n+\frac{1}{2}} = \frac{h}{2}\sum_{i=2}^{N-1}\left[\frac{\left(u_i^{n+1}\right)^2 + \left(u_i^n\right)^2}{2} - U\left(u_i^{n+1}\right) - U\left(u_i^n\right) + \frac{\beta}{2}\left(\frac{u_i^{n+1} - u_i^n}{\Delta\tau}\right)^2\right]$$

$$+ \frac{1}{2h}\sum_{i=1}^{N-1}\left\{\frac{\beta}{12}\left[\left(u_{i+1}^{n+1} - u_i^{n+1}\right)^2 + \left(u_{i+1}^n - u_i^n\right)^2\right] + \left(q_{i+1}^{n+\frac{1}{2}} - q_i^{n+\frac{1}{2}}\right)^2\right\}$$

are conserved by the difference scheme (1.10) in the sense that $M^{n+1} = M^n$ and $E^{n+1/2} = E^{n-1/2}$ for $n \geqslant 0$.

The above presented scheme and algorithm have been verified for different grids and time increments and the approximation has been confirmed.

Self-similar behavior. Results and discussion

Using the strongly implicit difference scheme with internal iterations, equation (1.10) allows us to follow the evolution of the solution at very long times. The system is rendered into a seven-diagonal band-matrix form and solved effectively by specialized solver with pivoting [17], which is stable to round-off errors even for 2 500 000 points of spatial resolution and up to 128 000 time steps. We conduct a few significant simulations, all of them with fixed parameters: $X_l = -20$, $X_r = 20$, $\beta_1 = 0.3$, $\beta_2 = 0.1$, $\alpha = 0.6$, $\Delta\tau = 0.05$, $h = 0.1$. Let us emphasize that a big set of investigations is implemented in [20, 21] concerning, however, only the supercritical modes. Here, we focus on the dynamical behavior of traveling localized solutions developing from critical initial data belonging to the subsonic and supersonic range of velocities. For phase velocities $c_l = 0.22$ and $c_r = -1.2$, the main solitary waves appear virtually non-deformed from the interaction, but additional oscillations are excited at the trailing edge of each one of them after the interaction (see figures 1.1–1.3). It is instructive to check whether this is an artifact of the difference scheme (due to numerical dissipation, instability of the difference scheme or other shortcomings) or it is an intrinsic property of the system under consideration. We extract the residual signals and track their evolution for very long times when they tend to adopt a self-similar shape (figure 1.4). Obviously their amplitudes decrease with the time τ approximately by the factor $\tau^{1/2}$, while the length scales increase by the same factor. Keeping in mind this observation, we test a hypothesis about the dependence on the time of the amplitude and the support of Airy-function shaped coherent structures described by function $\mathrm{Ai}(x)$, $x < 0$, as well as by the asymptotics of the Airy function for $x < 0$. Both give a very good quantitative agreement with the numerically obtained solutions (see figures 1.5 and 1.6). Let us remind you that the Airy function is defined by the improper integral

$$\mathrm{Ai}(x) = \frac{1}{\pi} \int_0^\infty \cos\left(\frac{t^3}{3} + xt\right) dt.$$

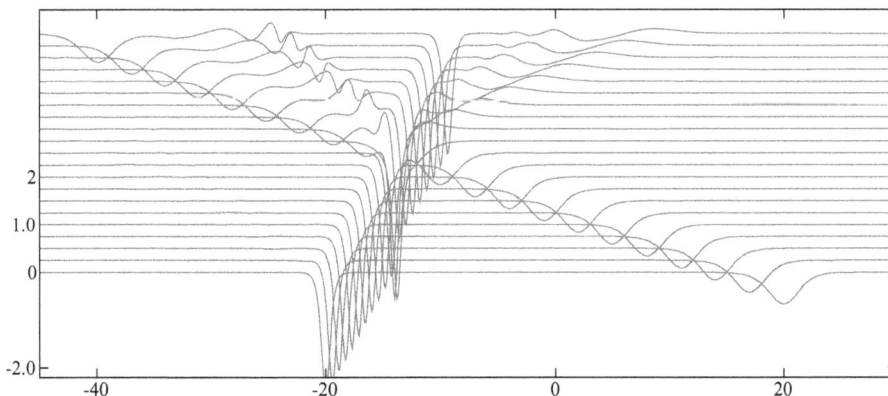

Figure 1.1. Head-on collision of depressions: subsonic $c_l = 0.22$, supersonic $c_r = -1.2$, $M_{tot} = -5.020\,02$, $E_{tot} = 3.584$. Short-times evolution. (Final time $\tau = 50$.) Forming of accompanying excitations after the interaction.

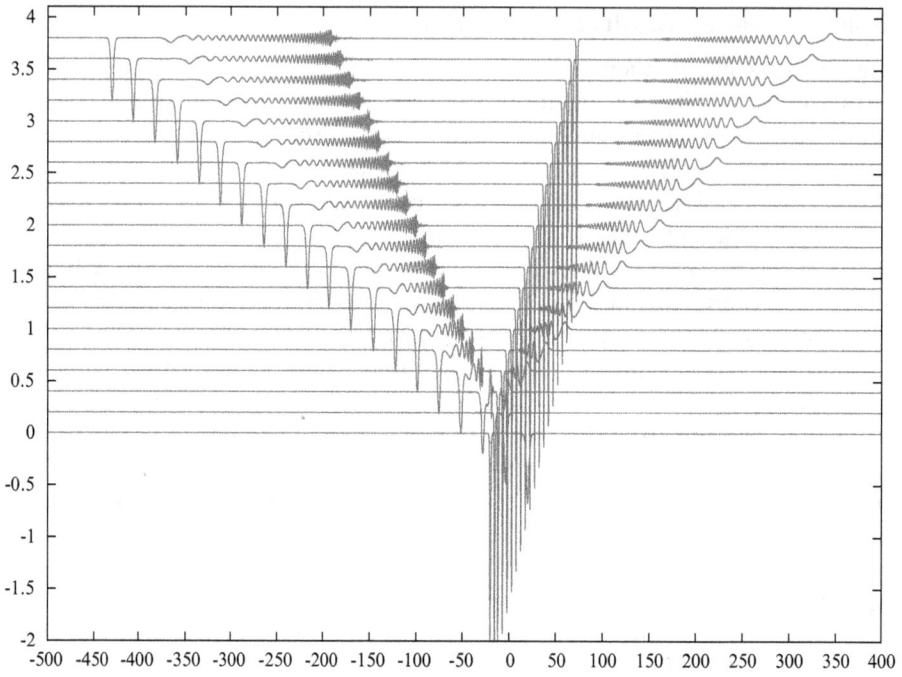

Figure 1.2. The same parameters as in figure 1.1. Middle-times evolution. (Final time $\tau = 420$, 8400 time steps $\Delta\tau = 0.05$.)

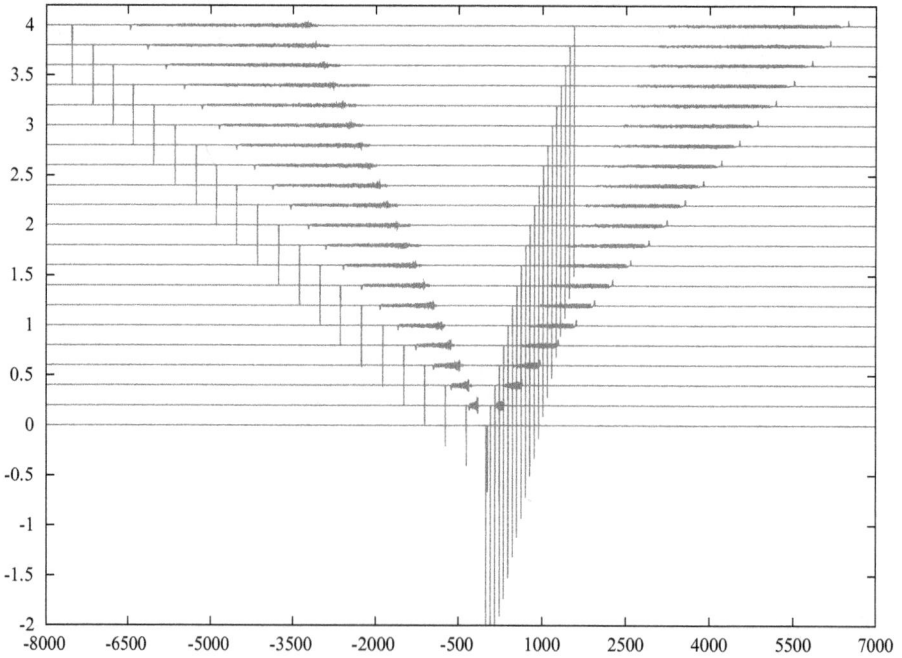

Figure 1.3. The same parameters as in figure 1.1. Very long-times evolution. (Final time $\tau = 6400$, 128 000 time steps $\Delta\tau = 0.05$.)

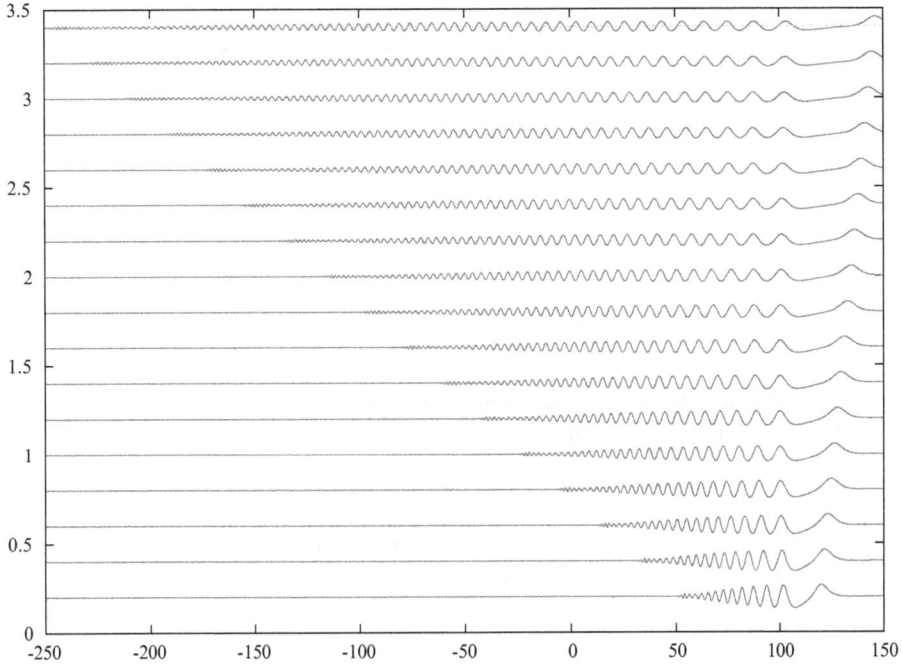

Figure 1.4. The time evolution of the left right-going trail (shifted) in the previous figures.

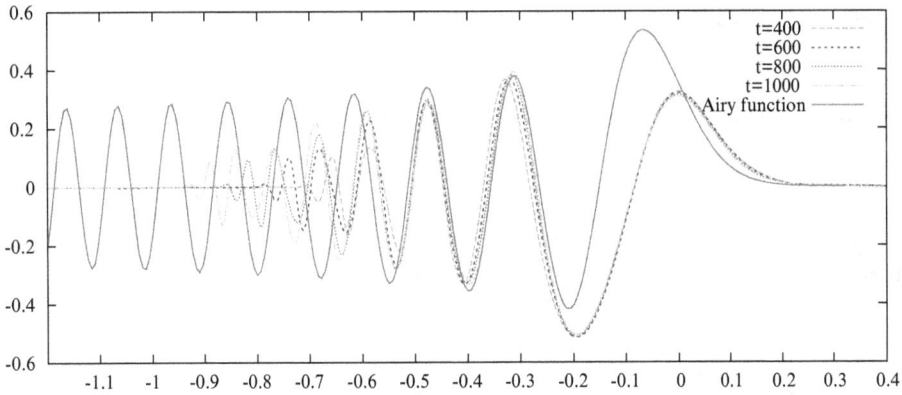

Figure 1.5. Scaled and shifted form of the left right-going trail for small and middle times compared and fitted to the Airy function.

It has a turning point at the origin oscillating in the negative part of x and decaying exponentially in the positive part of x. Also, the asymptotic behavior in the negative direction is

$$\mathrm{Ai}(-x) \sim \frac{\sin\left(\frac{2}{3}x^{\frac{3}{2}} + \frac{\pi}{4}\right)}{\sqrt{\pi}\, x^{\frac{1}{4}}}.$$

Yet, we calculate the individual masses and energies of these trails and establish that they are very small part of the total mass and the total energy not exceeding

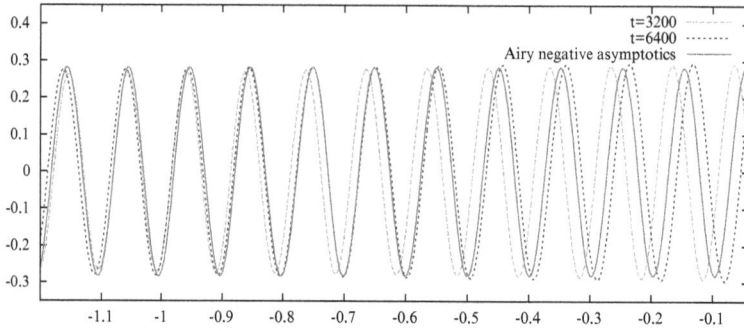

Figure 1.6. Scaled and shifted form of the left right-going trail for very long times and comparison with the Airy negative asymptotics ($x \rightarrow -\infty$).

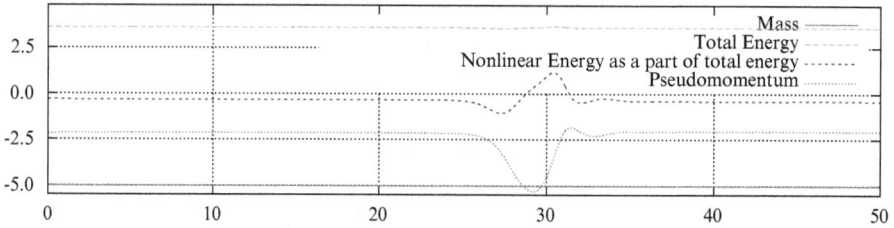

Figure 1.7. Head-on collision of the depressions in figures 1.1–1.3. Perfect conservation of the total mass $M = \text{const} = -5.02002$ and the total energy $E = \text{const} = 3.584$, the balance of pseudomomentum $\Delta P = 0.127$, $0 \leqslant \tau \leqslant 50$.

Table 1.1. The trails after the head-on collision of main solitons (depressions) from figures 1.1–1.3.

Pulses	M	E
Whole system	−5.02002	3.584
Right-going trail	−0.05404 (1.07% of M_{tot})	0.1052 (2.9% of E_{tot})
Left-going trail	−0.171 (3.4% of M_{tot})	0.15034 (4.2% of E_{tot})

3%–4% for each of them (see figure 1.7 and table 1.1). Moreover, we established that their individual masses and energies are perfectly conserved in the time. This is the reason to consider them as aggregate signals which are born after the soliton interaction which possess and propagate as independent self-similar structures not depending on the main solitons, which keep their quasi-particle behavior.

We conducted one more experiment—this time the subsonic velocity $c_1 = 0.4$ is close to the critical value $1/\sqrt{3}$. After the interaction, the slower soliton emanates one more highly oscillating self-similar trail with very small amplitude. We neglect its energy and mass because their magnitudes are very small and compatible with the order of approximation of the scheme. The main solitons keep their identity of quasi-particles after the interaction with insignificant phase shift (figures 1.8 and 1.9). The total mass and the total energy as well as individual ones are perfectly conserved (see table 1.2).

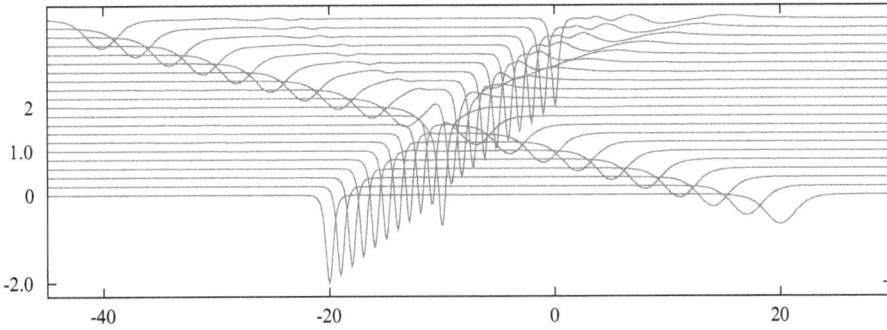

Figure 1.8. Head-on collision of depressions: subsonic $c_l = 0.4$, supersonic $c_r = -1.2$, $M_{tot} = -4.153$, $E_{tot} = 3.12076$. Short-times evolution. Forming of trails after the interaction.

Table 1.2. The trails after the head-on collision of main solitons (depressions) from figures 1.8 and 1.9.

Pulses	M	E
Whole system	−4.153	3.12076
Right-going trail	−0.05404 (1.3% of M_{tot})	0.1052 (3.3% of E_{tot})
Left-going trail	−0.02769 (0.7% of M_{tot})	0.00285 (0.1% of E_{tot})

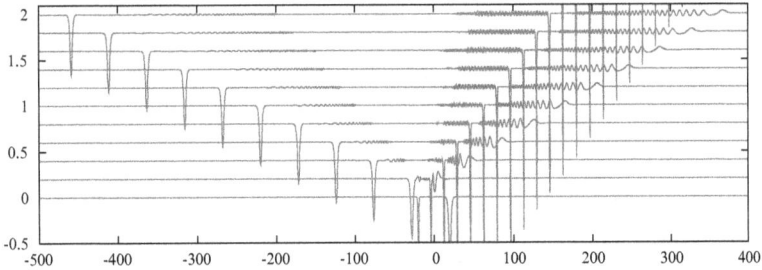

Figure 1.9. Middle-time evolution from figure 1.8.

The last simulation concerns another kind of solitons—so-called humps. We set again phase velocities to be from the both ranges, i.e. $c_l = 1.2$, $c_r = -0.22$ (figures 1.10–1.12). Due to the symmetry, the total mass $M_{tot} = 5.0202$ is positive but its magnitude coincides with those in the first example. The partition of the individual masses of the accompanying self-similar trails are the same as in the first example and the masses are positive. The energies total and individual are the same (see table 1.1).

Conclusions

We reach conclusions about the dynamics of interacting solitons (depressions and humps) of the dispersive wave system (DWS) as a one-dimensional Galilean

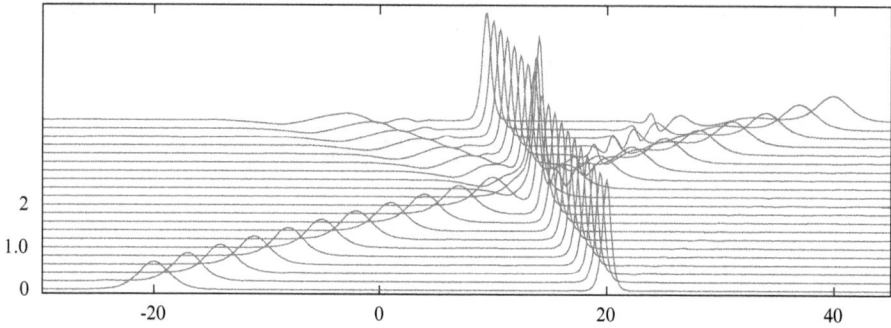

Figure 1.10. Head-on collision of humps: supersonic $c_l = -1.2$, subsonic $c_r = 0.22$, $M_{tot} = 5.020\ 02$, $E_{tot} = 3.584$. Short-times evolution. (Final time $\tau = 50$.) Forming of accompanying excitations.

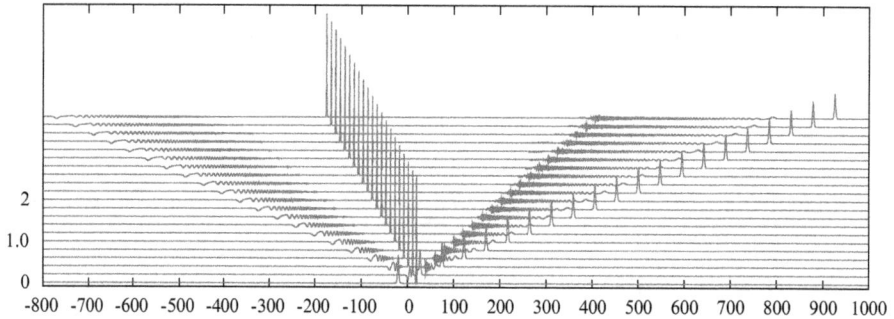

Figure 1.11. Middle-time evolution from figure 1.10.

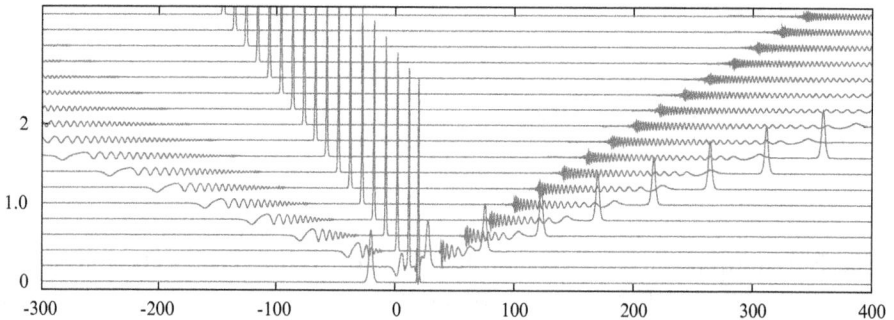

Figure 1.12. Middle-time evolution from figure 1.10. Magnification.

invariant of the energy-consistent Boussinesq system, investigated numerically by means of an energy conserving staggered difference scheme. The initial condition is constructed as a linear combination of the exact *sech*-solutions of the DWS for the shape of the stationary propagating solitons. Different combinations of solitons with subsonic and supersonic phase velocities are used as the initial condition and the evolution of the system is followed numerically for very long times (up to 128000 time steps). Using a mesh with 2500000 points of spatial resolution was a good test for the stability of the used difference scheme and enabled us to track the evolution of the main solitons as quasi-particles, as well as to observe and establish the self-similar behavior of the emanated trails. In our opinion, such a kind of effect is obtained for first time in a solution of DWS (though numerical) and it is a complement of the investigation conducted in [20, 21], where only configurations of supersonic solitons were conducted. A similar class of solutions adopting the self-similarity are observed in the coupled system of the generalized Boussinesq equation and the wave equation with the cubic root of time [27] and in the subsonic surface elevations (humps), considered as 'aging' soliton solutions of improved versions of the proper Boussinesq equation and the regularized long wave equation (see [32]). Also, such a self-similar solution can be found both for Burgers' equation [16] and the Korteweg–de Vries equation [44]. Similar natural effects are observed in the nonlinear optics. So, our results show that the soliton dynamics of DWS turns out to be very susceptible to the initial phase velocities of the solitons.

1.3 Numerical implementation of Fourier-transform method for generalized wave equations

Having in mind the difficulties related to resolving the nonlinear wave and dispersive dynamical systems we change the numerical approach replacing the difference schemes and finite-difference methods by pseudospectral method [47]. Transforming the original 2D Cauchy problem in Fourier space we reduce the spatial dimensions and get 1D Cauchy problem in configuration space. For linear problems it would be an excellent method. Here an obstacle are the nonlinear terms and their images in Fourier space. We get over this difficulty by introducing of internal iterations [32]— an original implementation of the classical Picard iteration for wave and dispersive equations. In this way we construct conservative finite-difference scheme for the linear part of the image of the operator in Fourier space and linearize the convolution integral by alternating the forward and backward Fourier transform. The method is tested and verified on Cauchy problems for 1D linear and nonlinear wave equations that admit analytic solutions. The successful implementation of the above method is a precondition for numerical solving of multidimensional dynamical systems by using of multidimensional Fourier transform.

Problem formulation

Consider the Boussinesq equation in two spatial dimensions (Boussinesq paradigm equation); equation (1.5). The initial conditions can be prepared by a single soliton (computed numerically and semi-analytically) or as a superposition of two solitons (see, for example, [10, 24, 25, 30]). The possible ways to numerically solve the above problem can be summarized in three groups: (i) by using a semi-implicit difference scheme; (ii) by using a fully implicit difference scheme; (iii) by using pseudospectral methods. In this paper, we focus our attention on the last ones.

Fourier integral-transform method

Instead of using a multigrid solver (see, for example, [31]), we can use a 2D Fourier transform. Applying it to the original equation (1.5), we get a second order ordinary differential equation (ODE) with respect to the time in the configurational space

$$[1 + 4\pi\beta_1(\xi^2 + \eta^2)]\hat{u}_{tt} = -4\pi^2(\xi^2 + \eta^2)[1 + 4\beta_2\pi^2(\xi^2 + \eta^2)]\hat{u}$$
$$+ 4\pi^2\alpha(\xi^2 + \eta^2)\hat{N} \tag{1.11}$$

where $\hat{u}(\xi, \eta, t) := \mathcal{F}[u]$ and $\hat{N}(\xi, \eta, t) := \mathcal{F}[u^2]$. Solving the last ODE is straightforward and requires very few operations per time step for given $\hat{N}$ but the lion's share of the computational resources are consumed by the computation of the contribution of the nonlinear term. An implicit scheme would require inverting the matrix that results from the discrete approximation of the convolution integral representing the Fourier transform of the nonlinear term u^2. The concept of the pseudospectral method is to use the inverse Fourier transform to represent the sought function in the configurational space and to compute the square there, and then to 'return' to the spectral space via the Fourier transform. The straightforward application of the pseudo-spectral method leads to an inherently explicit scheme, and in many cases, the latter is fully enough. Yet, for computations at very large times, one needs a fully conservative energy-conserving scheme. The latter is the object of the present investigation.

Numerical implementation of the pseudo-spectral method

We introduce a uniform grid (ξ_m, η_n) in the Fourier space and discretize the Fourier integral. Suppose that we know $\hat{u}^k$, $\hat{u}^{k-1}$, ..., $\hat{u}^0$. Then the next $(n + 1)$st time stage is computed from the following three-stage difference scheme

$$\left[1 + 4\pi\beta_1\left(\xi_m^2 + \eta_n^2\right)\right]\frac{\hat{u}_{mn}^{k,\,l+1} - 2\hat{u}_{mn}^k + \hat{u}_{mn}^{k-1}}{\tau^2}$$
$$= -2\pi^2\left(\xi_m^2 + \eta_n^2\right)\left[1 + 4\beta_2\pi^2\left(\xi_m^2 + \eta_n^2\right)\right]\left(\hat{u}_{mn}^{k,\,l+1} + \hat{u}_{mn}^{k-1}\right) \tag{1.12}$$
$$+ \frac{4}{3}\pi^2\alpha\left[\xi_m^2 + \eta_n^2\right]\mathcal{D}_\mathrm{F}\left[\left(\mathcal{D}_\mathrm{F}^{-1}\left[\hat{u}_{mn}^{k,\,l}\right]\right)^2 + \mathcal{D}_\mathrm{F}^{-1}\left[\hat{u}_{mn}^{k,\,l}\right]\mathcal{D}_\mathrm{F}^{-1}\left[\hat{u}_{mn}^{k-1}\right] + \left(\mathcal{D}_\mathrm{F}^{-1}\left[\hat{u}_{mn}^{k-1}\right]\right)^2\right],$$

where τ is the time step, $\mathcal{D}_\mathrm{F}[\cdot]$ denotes the discrete Fourier transform, and $\mathcal{D}_\mathrm{F}^{-1}[\cdot]$ is the inverse, respectively. The concept of internal iterations requires that at each time stage the linear scheme equation (1.12) starts with $u_{mn}^{k,\,l} = u_{mn}^k$, $l = 0$ and is repeated with increasing the number l until convergence is reached for some $l + 1 = L$. Then, it is set up that $u_{mn}^{k+1} := u_{mn}^{k,\,L}$. Then, following [32], we show that the scheme is fully nonlinear and fully implicit and conserves the energy within the tolerance level set for the convergence of the internal iterations (can be chosen close the round-off error of the computer). Note that the inverse Fourier transform gives a discrete function $u_{ij}^k := \mathcal{D}_F^{-1}[\hat{u}_{mn}^k]$, where i and j are the indices of a specific grid point in the configurational space.

Numerical tests and validation

We treat two 1D wave equations. In order to approximate the Fourier integrals, we use specialized Filon's quadrature [37] on a uniform mesh

$$\int_{-x_\infty}^{x_\infty} u(x) e^{i\xi x}\, dx \approx \left(\frac{1}{i\xi} + \frac{1 - e^{-i\xi h}}{\xi^2 h}\right) v_M - \left(\frac{1}{i\xi} + \frac{e^{-i\xi h} - 1}{\xi^2 h}\right) v_0 + \frac{4}{\xi^2 h} \sin^2\frac{\xi h}{2} \sum_{m=1}^{M-1} v_m,$$

with $v \equiv u(x)\, e^{i\xi x}$, spatial step h and 'actual' infinities $[-x_\infty, x_\infty]$.

The advantage of above quadrature consists of both—for $\xi h \leqslant 1$ it becomes a generalized trapezoidal formula with $O(h^2)$ error and when $\xi h > 1$ the order of error is as $O(M\xi^{-3} u_{xx})$ [33, 38]. Keeping in mind the localized nature of the sought solutions, it is obvious that $\lim_{x \to \pm\infty} u_{xx} = 0$ and the decay of the quadrature error for $\xi \gg 1$ and given x_∞ in the problems in question is obeyed.

Cauchy problem for 1D string equation

Let us consider the well-known Cauchy problem

$$u_{tt} = c^2 u_{xx}, \qquad c = \text{const} > 0, \qquad -\infty < x < \infty, \qquad t > 0 \tag{1.13}$$

$$u(x, 0) = f(x), \qquad u_t(x, 0) = g(x) \tag{1.14}$$

with the exact solution given by D'Alembert's formula

$$u(x, t) = \frac{1}{2}[f(x - ct) + f(x + ct)] + \frac{1}{2c} \int_{x-ct}^{x+ct} g\, d\theta.$$

The image of the problem (1.13) and (1.14) in the configurational space (again Cauchy problem with respect an ODE with algebraic right hand side) reads

$$\hat{u}_{tt} = -c^2 \xi^2 \hat{u}, \qquad \hat{u}(\xi, 0) = \hat{f}(\xi) \qquad \hat{u}_t(\xi, 0) = \hat{g}(\xi) \tag{1.15}$$

and the exact solution $\hat{u}(\xi, t) = \hat{f}(\xi) \cos c\xi t + \frac{\hat{g}(\xi)}{c\xi} \sin c\xi t$ where $\mathcal{F}^{-1}[\hat{u}] = u(x, t)$.

Following the idea in (1.12), we build a standard three-stage explicit difference scheme for (1.15)

$$\frac{\hat{u}_m^{k+1} - 2\hat{u}_m^k + \hat{u}_m^{k-1}}{\tau^2} = -\frac{c^2 \xi^2}{2}(\hat{u}_m^{k+1} + \hat{u}_m^{k-1}) \tag{1.16}$$

setting the phase velocity $c = 1$, and (i) $f(x) = e^{-(x-X)^2}$, $g(x) = 2(x - X)e^{-(x-X)^2}$, X stands for the initial position of the center of the solitary wave; (ii) the functions $f(x)$ and $g(x)$ in the initial conditions are sech-like (see the next subsection). Let us note that the scheme is stable when $ch/\tau \leqslant 1$.

Regularized long wave equation

If $\beta_1 = 0$, the Boussinesq equation reduces to the RLWE, equation (1.1), and possesses the following exact solitary-wave solution (see [32]):

$$w = -\frac{3}{2}\frac{c^2 - 1}{\alpha}\,\mathrm{sech}^2\!\left(\frac{x - ct}{2c}\sqrt{\frac{c^2 - 1}{\beta}}\right). \tag{1.17}$$

Here, c is the phase velocity, α is the parameter of the nonlinearity, and β is the dispersion parameter. For the mechanical meaning of equation (1.1), we refer the reader to [32]. To begin the time stepping, we set

$$u(x, 0) = w(x, 0) \quad \text{and} \quad u(x, \tau) = w_t(x, 0)\tau + w(x, 0) \tag{1.18}$$

and transform the latter to spectral space, thus providing the two initial conditions for the 1D version of the scheme (1.12).

Some results and discussion

We show and discuss two groups of results concerning the 1D linear string equation and the 1D nonlinear RLWE. Figure 1.13 demonstrates the excellent comparison between the D'Alembert solution (dashed lines) and the numerical solution by the pseudospectral method (solid lines). Two running waves with Gaussian shape start from the coordinate origin $X = 0$ and go unchanged to the left and to the right with phase velocities $c_l = -c_r = 1$. The conclusion is that the linear wave equations can be discretized and solved numerically in the spectral space and only after the solution is obtained at each time stage, the inverse Fourier transformation can be used to restore the solution in the configuration space. As a rule, the mapped differential equations are simpler compared to the original ones.

In figure 1.14, the wave shapes are the same but the initial condition is a superposition of two running waves starting from different positions $-X_l = X_r = 3.5$ again with phase velocities $c_l = -c_r = 1$ which collide between them elastically.

The second part of investigation concerns 1D nonlinear dispersive generalized wave equations using RLWE as a featuring example. In the following figures, the obtained numerical solutions with the here described algorithm are presented. To test the reliability of the method, we compare the obtained results with those obtained by a finite difference method in [32].

In figures 1.15 and 1.16, the head-on collisions for supercritical phase speeds that are still below the threshold of the blow-up are presented. The first figure

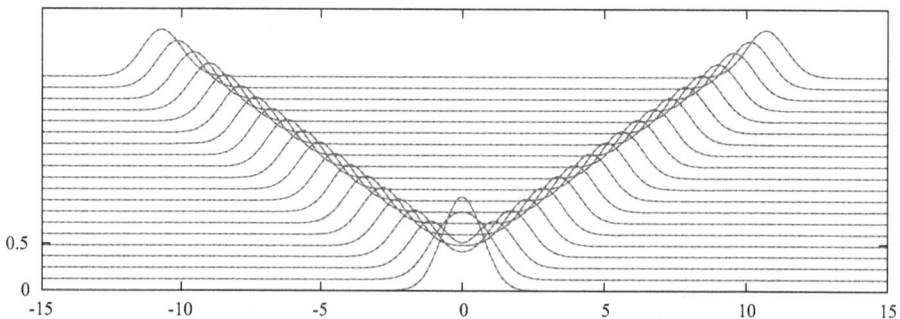

Figure 1.13. Comparison of the numerical solution with the D'Alembert formula.

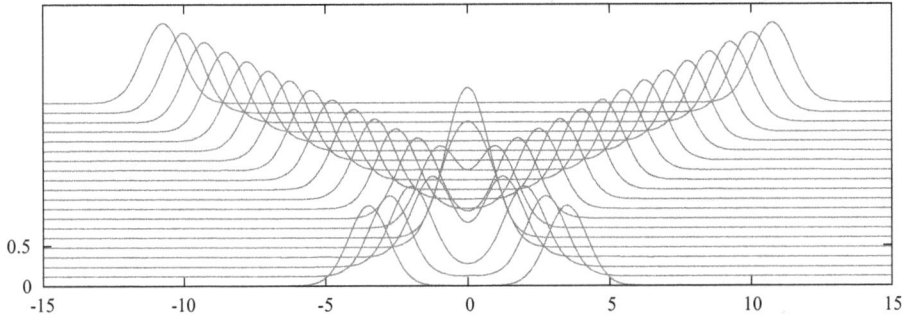

Figure 1.14. Superposition and elastic interaction of two Gaussian pulses.

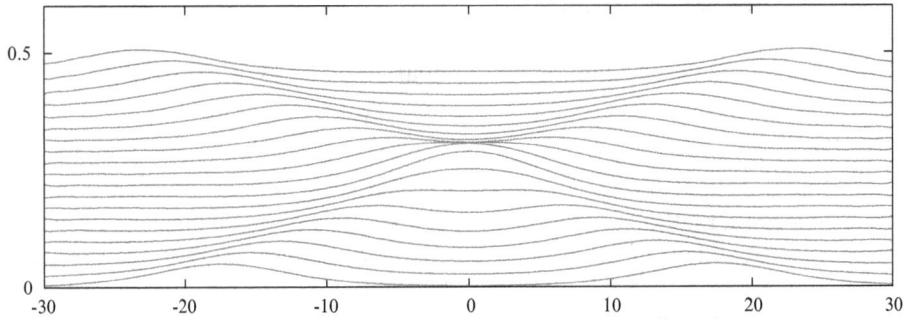

Figure 1.15. The inelastic interaction in RLWE for slightly supercritical phase velocities, $c_l = -c_r = 1.05$, $\alpha = -3$, $\beta = 1$.

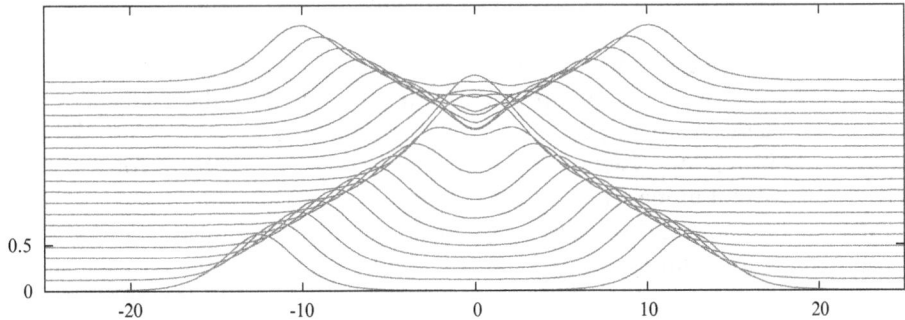

Figure 1.16. The interaction in RLWE near to the threshold of nonlinear blow-up, $c_l = -c_r = 1.5$, $\alpha = -3$, $\beta = 1$.

presents a case where the nonlinearity is weaker, while in the second of these figures, the nonlinearity is considerable. In both cases, the solitons retain their individualities after the collision and no significant radiation is observed despite the fact that RLWE is not a fully integrable case. The only sign of inelasticity is the phase shift experienced by the colliding waves. For the sake of saving space, it is not presented here.

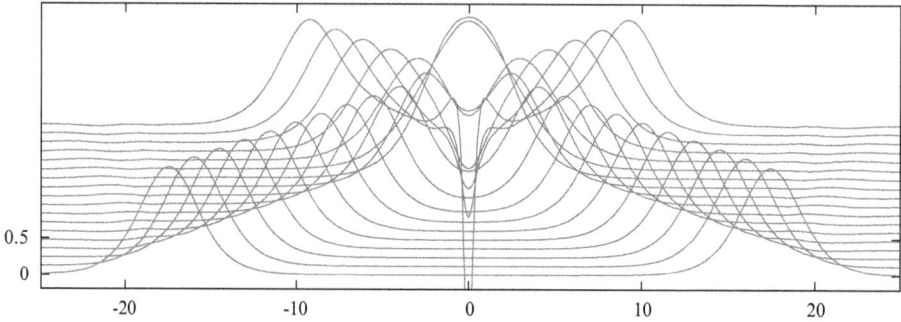

Figure 1.17. The blow-up in RLWE for large supercritical phase velocities, $c_1 = -c_r = 2$, $\alpha = -3$, $\beta = 1$.

In the end, we present in figure 1.17 a case known to lead to a blow-up of the solution.

In all considered cases, an excellent comparison with [32] is observed.

Conclusion

Though this investigation has a preliminary character, we have demonstrated that the pseudospectral methods and in particular Fourier transform can be efficient both for numerical treatment of linear and nonlinear wave equations. For the 2D and 3D equations, one needs to apply 2D and 3D Fourier transforms and to follow the procedures described above.

1.4 Perturbation solution for the 2D shallow-water waves

In [30], we used a perturbation approach to build an approximate initial condition for the Cauchy problem about a modified 2D Boussinesq paradigm equation. This is a shallow water model, which contains nonlinearity and fourth-order dispersion. Considering $\varepsilon = c^2$ as a small parameter (c—the phase speed of the localized wave), a perturbation series with respect to the small parameter $\varepsilon := c^2$ is developed. A hierarchical bifurcation system of ordinary differential equations of fourth order is derived. We solve it numerically and the obtained result is used to shape a 2D initial condition. The numerical scheme uses a special approximation for the behavioral condition in the singularity point (the origin). Its shape being a solitary wave decades algebraically at the infinity rather than exponentially as in the 1D cases. The new result can be instrumental for understanding the interaction of 2D Boussinesq solitons, and for creating more efficient numerical algorithms explicitly acknowledging the asymptotic behavior of the solution.

We term the equation in question the cubic-quintic Boussinesq paradigm equation (CQBPE)

$$u_{tt} = \Delta[u - \alpha(u^3 - \sigma u^5) + \beta_1 u_{tt} - \beta_2 \Delta u] \tag{1.19}$$

where u is again surface elevation, β_1, where $\beta_1 > 0$ are dispersion coefficients. Unlike the 2D Boussinesq paradigm with quadratic nonlinearity, equation (1.5), this equation has a positive energy (see below) and hence a limited amplitude for long-time evolution. The latter is due to the presence of the nonlinearity of fifth degree and the introduced parameter σ.

Preliminaries

After the resounding success of the quasi-particle (QP) concept in 1D, one may have expected that the attention would have turned in full swing to multidimensions. Unfortunately, for the time being, the 2D Boussinesq model is still less amenable to analytical techniques. The ultimate goal is to find the collision properties of the 2D localized solutions, but even the existence of 2D stationary-propagating localized solution cannot be established numerically. This requires the development of numerical techniques. One of the main difficulties for the difference schemes lies in the inevitable reducing of the infinite interval to a finite one. This can be surmounted if a spectral method is used with a basis system of localized functions, which automatically acknowledge the requirement that the solution belongs to $L^2(-\infty, \infty)$ space. Along these lines, a specialized Galerkin spectral technique was proposed in [14], and applied to the 2D stationary propagating 2D Boussinesq wave in [12, 13]. A perturbation technique for slow to moderate phase speed has been elaborated in [24]. A special kind of boundary condition of the type of the perfectly matched layer was used in [22] and the results were confirmed with high accuracy by a difference scheme.

The profiles obtained for the stationary propagating QP were used for a first time for time evolution of the solution in [10], where an innovative idea (see [41] and the literature cited therein) for a Riemann-like solver was transferred to the case under consideration. The time evolution as computed in [10] showed that for phase speeds $c < 0.3$, the shape moved steadily until approximately 12 dimensionless time units and then dispersed under the influence of the higher-order derivatives. Respectively, if the phase speed exceeds the critical value $c = 0.3$, the nonlinearity dominates, but in this case, the solution blows up. In order to ensure that this is not a numerical artifact, we constructed a difference scheme and verified the findings of the pioneering work with high accuracy (see [25]). This means that if the cause of blow-up is removed, we can expect to extend the interval in which the solitary wave behaves as a QP, if we are able to solve for higher phase speeds.

In 1D, it was shown in [48, 49] that the quadratic nonlinearity of the equation is the cause of the blow up. The appearance of blow up was confirmed in the numerical investigations [32]. We have embarked on providing the analytical tools for obtaining the estimates in 2D (see the work in the present volume [42]). Yet, it is important to modify the nonlinearity of the Boussinesq equations in a manner where the blow-up can be avoided. We take a clue of the work [43] where cubic and quintic terms were shown to replace the quadratic nonlinearity in some models with significance in theory of atomic chains. Later on, it was shown numerically in [26] that for a very wide range of phase speeds, no blow up occurs. Guided by this example, we consider a Boussinesq equation with cubic and quintic nonlinear terms and use the perturbation technique to find the stationary propagating solitary wave of this model.

The cubic-quintic Boussinesq paradigm equation (CQBPE)

We focus here on the two-dimensional amplitude equation, CQBPE, equation (1.19). It is easily shown that one can introduce new spatial and temporal coordinates for which one of the dispersion coefficients can be made equal to unity. We prefer to have $\beta_2 = 1$, because we will consider only the full fledged paradigm equations when $\beta_2 \neq 0$. The parameter σ accounts for the relative importance of the quintic nonlinearity term. A note on the notation: in the original BE as related to the water waves, the nonlinear term has a positive sign, and the solutions are actually depressions for the subcritical case. Here, we have deliberately changed the sign for the sake of the presentation.

Since the amplitude parameter α can be eliminated by rescaling the solution with simultaneous change of σ, without loosing the generality, we will consider here only the case $\alpha = 1$. In order to derive the energy law for equation (1.19), we rewrite it as the following system

$$u_t = \Delta q, \quad q_t = u - (u^3 - \sigma u^5) + \beta_1 u_{tt} - \beta_2 \Delta u. \tag{1.20}$$

Then, the energy law reads

$$\frac{dE}{dt} = 0,$$

$$E = \frac{1}{2} \int_{-\infty}^{\infty} \int_{-\infty}^{\infty} \left[(\nabla q)^2 + u^2 - \frac{1}{2}u^4 + \frac{\sigma}{3}u^6 + \beta_1 u^2 + \beta_2 (\nabla u)^2 \right] dx \, dy. \tag{1.21}$$

Unlike the BPE with quadratic nonlinearity, when the amplitude increases, the quintic term in CQBPE for reasonably large σ will dominate and will make the energy functional positive, which limits the increase of the amplitude. All this means that no blow-up can be expected for CQBPE if the bi-quadratic form $1 - \frac{1}{2}u^2 + \frac{1}{3}\sigma u^4$ is always positive when $\sigma > \frac{3}{16}$. However, this case is clearly of no interest, because with the strictly positive-definite energy, the existence of bifurcation and the appearance of a nontrivial solution-like solution may not be possible. Let us concentrate our attention on the case $0 < \sigma \leqslant \frac{3}{16}$. Then, for smaller u, the term in the energy containing only u can become even negative and the balance with the positive terms containing the gradients can define a permanent structure (solitary wave). In the above interval for σ, the bi-quadratic form adopts the form

$$1 - \frac{1}{2}u^2 + \frac{1}{3}\sigma u^4 = \frac{1}{3}\sigma[(u^2 - u_1)(u^2 - u_2)], \quad \text{where } u_{1,2} := \frac{3}{4}\sigma^{-1}\left(1 \pm \sqrt{1 - \frac{16}{3}\sigma}\right).$$

It is clear that the last term can have negative values for $u_2 < u^2 < u_1$. This means that, in the spatial intervals where this happens, it will act to destabilize the trivial solution. Yet, a blow-up cannot be reached because immediately after u^2 becomes greater than u_1, the complete positive definiteness of the energy functional is restored.

Note that the blow-up occurs when the negative term can start increasing in time. This happens when the amplitude of the function u increases.

To find a stationary moving solitary wave, we introduce relative coordinates $\hat{x} = x - c_1 t$, $\hat{y} = y - c_2 t$, in a frame moving with velocity (c_1, c_2). Since there is no evolution in the moving frame $v(x, y, t) = u(\hat{x}, \hat{y})$, the following equation holds for u:

$$
\begin{aligned}
\left(c_1^2 u_{\hat{x}\hat{x}} + 2c_1 c_2 u_{\hat{x}\hat{y}} + c_2^2 u_{\hat{y}\hat{y}}\right) &= (u_{\hat{x}\hat{x}} + u_{\hat{y}\hat{y}}) \\
&\quad - [(u^3 - \sigma u^5)\hat{x}\hat{x} + (u^3 - \sigma u^5)\hat{y}\hat{y})] \\
&\quad - (u_{\hat{x}\hat{x}\hat{x}\hat{x}} + 2u_{\hat{x}\hat{x}\hat{y}\hat{y}} + u_{\hat{y}\hat{y}\hat{y}\hat{y}}) \\
&\quad + \beta_1 \big[c_1^2 (u_{\hat{x}\hat{x}\hat{x}\hat{x}} + u_{\hat{x}\hat{x}\hat{y}\hat{y}}) + 2c_1 c_2 (u_{\hat{x}\hat{x}\hat{x}\hat{y}} + u_{\hat{x}\hat{y}\hat{y}\hat{y}}) \\
&\quad + c_2^2 (u_{\hat{x}\hat{x}\hat{y}\hat{y}} + u_{\hat{y}\hat{y}\hat{y}\hat{y}}) \big].
\end{aligned}
\tag{1.22}
$$

The so-called asymptotic boundary conditions (a.b.c.) read $u \to 0$, for $\hat{x} \to \pm\infty$, $\hat{y} \to \pm\infty$. The a.b.c.'s are invariant under rotation of the coordinate system, hence it is enough to consider solitary propagating wave along one of the coordinate axes, only. We chose $c_1 = 0$, $c_2 = c \neq 0$. Without fear of confusion, we will 'reset' the names of the independent variables to x, y and omit in what follows the hat over the function u.

Perturbation method

We follow [24] and create an asymptotic solution valid for small phase speeds of the soliton. We set the amplitude parameter $\alpha = 1$, because it can always be eliminated by rescaling the solution. We can also select $\beta_2 = 1$. This leaves us with only one parameter, β_1, apart from the phase speed c.

The small parameter does not multiply the highest derivative, hence the expansion is regular (see, e.g. [34]). When $c = 0$, the solution possesses a radial symmetry, and we consider the expansion

$$
u(x, y) = u_0(r) + \varepsilon u_1(x, y) + \varepsilon^2 u_2(x, y) + O(\varepsilon^3), \quad r = \sqrt{x^2 + y^2}
\tag{1.23}
$$

for which we obtain the following equations

$$
(u_0 + \varepsilon u_1 + \varepsilon^2 u_2)^3 \approx u_0^3 + 3\varepsilon u_0 u_1^2 + 3\varepsilon^2 (u_0 u_1^2 + u_0^2 u_2) + O(\varepsilon^3),
\tag{1.24}
$$

$$
(u_0 + \varepsilon u_1 + \varepsilon^2 u_2)^5 \approx u_0^5 + 5\varepsilon u_0^4 u_1 + 10\varepsilon^2 u_0^3 u_1^2 + 5\varepsilon^2 u_0^4 u_2 + O(\varepsilon^3).
\tag{1.25}
$$

Now, neglecting the terms of order $O(\varepsilon^3)$, we get for the three lowest orders in ε the following system

$$
\frac{1}{r}\frac{d}{dr}r\frac{d}{dr}\left[u_0(r) - u_0^3(r) + \sigma u_0^5(r) - \frac{1}{r}\frac{d}{dr}r\frac{du_0}{dr}\right] = 0,
\tag{1.26a}
$$

$$
\varepsilon\left[-\frac{d^2}{dy^2}u_0 + \beta_1\frac{d^2}{dy^2}\Delta u_0 + \Delta u_1 - 3\Delta(u_0^2 u_1) + 5\sigma\Delta(u_0^4 u_1) - \Delta^2 u_1\right] = 0,
\tag{1.26b}
$$

$$\varepsilon^2 \left[-\frac{d^2}{dy^2}u_1 + \beta_1\frac{d^2}{dy^2}\Delta u_1 + \Delta u_2 - 3\Delta\left(u_0u_1^2 + u_0^2u_2\right) \right.$$

$$\left. + 10\sigma\Delta\left(u_0^3u_1^2\right) + 5\sigma\Delta\left(u_0^4u_2\right) - 2\Delta(u_0u_2) - \Delta^2u_2 \right] = 0.$$

(1.26c)

We prefer to treat the above system in polar coordinates because then the region is unbounded only with respect to one of the variables (the polar radius r). The connection between the Cartesian and polar coordinates is given by $x = r\cos(\theta)$, $y = r\sin(\theta)$, where θ is the polar angle. Now for the derivative with respect to y, we have

$$\frac{\partial^2}{\partial y^2} := \sin^2\theta\frac{\partial^2}{\partial r^2} + \frac{\cos^2\theta}{r^2}\frac{\partial^2}{\partial\theta^2} + \frac{\sin 2\theta}{r}\frac{\partial^2}{\partial r\partial\theta} - \frac{\sin 2\theta}{r^2}\frac{\partial}{\partial\theta} + \frac{\cos^2\theta}{r}\frac{\partial}{\partial r}. \quad (1.27)$$

Applying this operator twice, we find the expression for the fourth derivative with respect to y. The Laplace operator in polar coordinates is well known and is omitted here.

We denote $F(r) = u_0(r)$. When manipulating the equation for u_1, we observe that the operator in equation (1.27) has to be applied only to the function u_0, which is independent of the polar angle θ and we get the following

$$\frac{\partial^2}{\partial y^2}u_0(r) = \sin^2\theta\frac{d^2F}{dr^2} + \frac{\cos^2\theta}{r}\frac{dF}{dr} = \frac{1}{2}\left(\frac{d^2F}{dr^2} + \frac{1}{r}\frac{dF}{dr}\right) + \frac{1}{2}\left(-\frac{d^2F}{dr^2} + \frac{1}{r}\frac{dF}{dr}\right)\cos(2\theta),$$

$$\frac{\partial^2}{\partial y^2}\Delta u_0(r) = \sin^2\theta\frac{d^2P}{dr^2} + \frac{\cos^2\theta}{r}\frac{dP}{dr} = \frac{1}{2}\left(\frac{d^2P}{dr^2} + \frac{1}{r}\frac{dP}{dr}\right) + \frac{1}{2}\left(-\frac{d^2P}{dr^2} + \frac{1}{r}\frac{dP}{dr}\right)\cos(2\theta),$$

$$P(r) := \frac{1}{r}\frac{d}{dt}r\frac{1}{r}\frac{d}{dt}F(r).$$

We limit ourselves to the $O(\epsilon)$ approximation, because it is usually enough for the purposes of an initial condition. Then, we can recast equation (1.26b) as follows

$$\left(\frac{1}{r}\frac{d}{dr}r\frac{d}{dr} + \frac{1}{r^2}\frac{\partial^2}{\partial\theta^2}\right)\left[u_1(r, \theta) - 3F^2(r)u_1(r, \theta) + 5F^4(r)u_1(r, \theta)\right.$$

$$\left. -\left(\frac{1}{r}\frac{d}{dr}r\frac{d}{dr} + \frac{1}{r^2}\frac{\partial^2}{\partial\theta^2}\right)u_1(r, \theta)\right]$$

(1.28)

$$= \sin^2\theta\frac{d^2}{dr^2}F(r) + \frac{\cos^2\theta}{r}\frac{d}{dr}F(r) - \beta_1\left[\sin^2\theta\frac{d^2}{dr^2}P(r) + \frac{\cos^2\theta}{r}\frac{d}{dr}P(r)\right].$$

The form of the right-hand side of equation (1.28) suggests that the following type of solution can be found

$$u_1(r, \theta) = G(r) - \beta_1Q(r) + [H(r) - \beta_1R(r)]\cos(2\theta). \quad (1.29)$$

The governing system

Before proceeding with the derivation of the system for the functions F, G, H and P, Q, R, we observe that the higher-order Bessel operators involved in those equations have the form (see [24]):

$$\frac{d^2}{dr^2} + \frac{1}{r}\frac{d}{dr} - \frac{4}{r^2} \equiv r\frac{d}{dr}\frac{1}{r^3}\frac{d}{dr}r^2, \qquad \frac{d^2}{dr^2} + \frac{1}{r}\frac{d}{dr} - \frac{16}{r^2} \equiv r^3\frac{d}{dr}\frac{1}{r^7}\frac{d}{dr}r^4. \quad (1.30)$$

We can integrate equation (1.26a) and set the two integration constants equal to zero. Thus, we get a lower-order equation for $F(r)$:

$$F(r) - F^3(r) + \sigma F^5(r) - \frac{1}{r}\frac{d}{dr}r\frac{dF}{dr} = 0. \quad (1.31a)$$

Upon substituting equation (1.29) into equation (1.28), and integrating under the asymptotic boundary conditions twice, we get the following equations for G and Q:

$$-G(r) + 3F^2(r)G(r) - 5\sigma F^4(r)G(r) + \frac{d^2G}{dr^2} + \frac{1}{r}\frac{dG}{dr} = -\frac{1}{2}F(r), \quad (1.31b)$$

$$-Q(r) + 3F^2(r)Q(r) - 5\sigma F^4(r)Q(r) + \frac{d^2Q}{dr^2} + \frac{1}{r}\frac{dQ}{dr} = -\frac{1}{2}P(r). \quad (1.31c)$$

Consistently implementing the above idea (grouping the terms with the same dependence on θ), we get the equations for the other functions, H, R, namely:

$$r\frac{d}{dr}\frac{1}{r^3}\frac{d}{dr}r^2\left[-H(r) + 3F^2(r)H(r) - 5\sigma F^4(r)H(r) + r\frac{d}{dr}\frac{1}{r^3}\frac{d}{dr}r^2H(r)\right]$$
$$= \frac{1}{2}\left[\frac{d^2}{dr^2}F(r) - \frac{1}{r}\frac{d}{dr}F(r)\right], \quad (1.31d)$$

$$r\frac{d}{dr}\frac{1}{r^3}\frac{d}{dr}r^2\left[-R(r) + 3F^2(r)R(r) - 5\sigma F^4(r)R(r) + r\frac{d}{dr}\frac{1}{r^3}\frac{d}{dr}r^2R(r)\right]$$
$$= \frac{1}{2}\left[\frac{d^2}{dr^2}P(r) - \frac{1}{r}\frac{d}{dr}P(r)\right]. \quad (1.31e)$$

When one is faced with singularities that arise from the use of specific coordinates (e.g. polar coordinates), he has to ensure the proper behavior of the functions in the point of singularity by imposing additional (purely mathematical) conditions in the geometric singularity called 'behavioral' (see [9]). The behavioral conditions at the origin arise from the fact that there is a singularity in the operator:

$$G'(0) = G'''(0) = Q'(0) = Q'''(0) = H'(0) = H'''(0) = R'(0) = R'''(0) = 0, \quad (1.32)$$

while the behavioral conditions at infinity (called asymptotic boundary conditions or a.b.c.) are given by

$$G(r),\ Q(r),\ H(r),\ R(r) \to 0 \quad \text{for} \quad r \to \infty.$$

The equations possess non-trivial solutions provided a nontrivial solution where $F(r) \not\equiv 0$ is found, because the rest of the equations are linear. Thus, one can tackle the bifurcation problem while finding the function $F(r)$.

Results

For the very elaborate technical details of the difference schemes for the above equations, we refer the reader to [24]. Respectively, the details of the application of the spectral methods can be found for the 1D case in [23] and for 2D case in [12, 15]. For the lack of space, we do not present here the verifications of the present results by means of the spectral algorithm from [12]. In figure 1.18, we present the results for the three sought functions. The right panel shows the results for F, G and H in linear, log–linear and log–log scales, alongside the comparison to the best-fit approximations. It is clearly seen that the asymptotic decay for H is indeed proportional to the inverse square of the polar coordinate. The results also testify that the 'computational infinity' $r_\infty = 1000$ is fully adequate for finding $H(r)$ with good accuracy. Note that because of the exponential decay of F and G, we augment the needed values of their grid versions with zeros in the region $50 < r < 1000$. Respectively, figure 1.19 gives the result for the functions corresponding to terms that are multiplied by β_1.

Since the primary purpose of this short note is to create an initial condition for time stepping algorithms, we need to present the results as best-fit analytic approximations for the functions F, G, H. Essentially, this is the same approach as in [24], but now we are more restricted because the new time stepping algorithm developed in [25] also requires initial conditions for the Laplacian of the functions. This prevents us from using first powers in the approximation. We found the following best-fit approximations for the shape of the stationary propagating

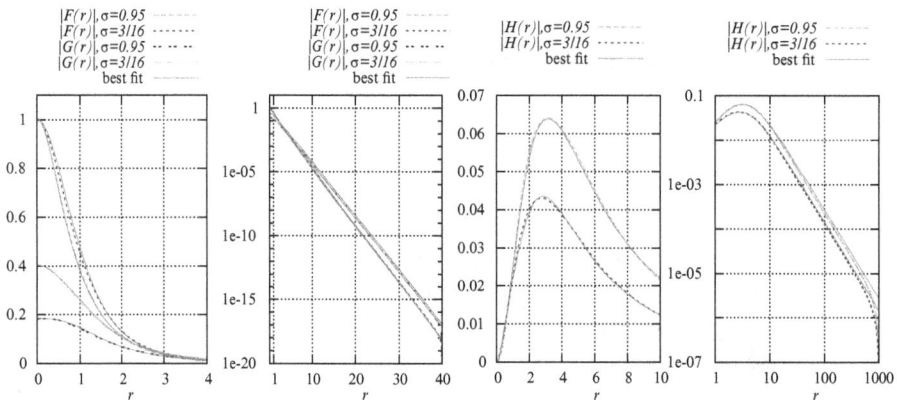

Figure 1.18. Finite-difference solutions for the $F(r)$, $G(r)$, $H(r)$ for two values of $\sigma = 0.95$ and $\sigma = 3/16$. The left panels present the behavior near the origin and the right panels present the behavior in the far fields in logarithmic coordinates. All panels give the best-fit functions denoted further as $f(r)$, $g(r)$ and $h(r)$ whose analytical representations are shown in the tables 1.3 and 1.4, rows 1, 2 and 3.

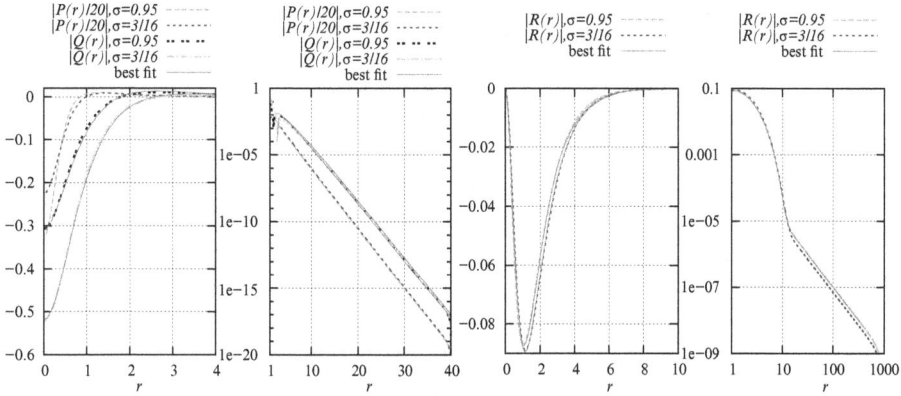

Figure 1.19. Finite-difference solutions for the $P(r)$, $Q(r)$, $R(r)$ for two values of $\sigma = 0.95$ and $\sigma = 3/16$. The left panels present the behavior near the origin and the right panels present the behavior in the far fields in logarithmic coordinates. All panels give the corresponding best-fit functions $\hat{f}(r)$, $\hat{g}(r)$ and $\hat{h}(r)$, whose analytical expressions are presented in the tables 1.3 and 1.4, rows 4 and 5.

Table 1.3. Best-fit functions depending on the weight coefficient $\sigma = 0.95$.

$f(x, y)$	$1.0032\dfrac{1 + 0.32\, r^2}{(1 + 1.86\, r^2 + 0.032\, r^4)^{0.75}}\operatorname{sech}(r)$
$g(x, y)$	$0.203(1 + 0.1\, r^2)^{0.25}[1.2\operatorname{sech}(r) - 0.3\operatorname{sech}(2r)]$
$h(x, y)$	$\dfrac{0.05 r^2 + 0.071 r^4}{1 + 3.2 r^2 + 0.6 r^4 + 0.026 r^6}$
$\hat{g}(x, y)$	$(1 + 3r^2)^{0.25}[0.085\operatorname{sech}(r) - 0.44\operatorname{sech}(1.86r) + 0.102\operatorname{sech}(2.2r) - 0.056\operatorname{sech}(4r)]$
$\hat{h}(x, y)$	$\dfrac{-0.36 r^2}{(1 + r^2)^2}[0.87\operatorname{sech}(2r) + 0.02 r^2\operatorname{sech}(1.51r) - 0.001]$

solitons for $\beta_2 = 1$ and two concrete values of the weight coefficient σ (see tables 1.3 and 1.4), namely:

$$u^s(x, y, t; c) = f(x, y) + c^2[g(x, y) + h(x, y)\cos(2\theta)] \\ - \beta_1 c^2[\hat{g}(x, y) + \hat{h}(x, y)\cos(2\theta)],$$

(1.33)

where $\theta(x, y) = \arctan(y/x)$.

We need to make sure here that the Laplacians of our functions F, G, and H are also smooth and well behaved, in order to be able to use them in algorithms of the type of [25]. In figure 1.20, the Laplacians of the three main functions F, G and H are plotted. The results for Q and R are very similar.

In the end, we present in figure 1.21 the shape of the solitary wave and the contour lines of the cross-sections for four different combinations of parameters.

It is observed that for the nontrivial value of the phase speed (specifically $c = 0.6$), the shape undergoes significant contraction in the direction of the propagation. In order to demonstrate this, we use a special set of non-uniformly distributed values

Table 1.4. Best-fit functions depending on the weight coefficient $\sigma = \frac{3}{16}$.

$f(x, y)$	$1.0032\dfrac{1 + 0.31\,r^2}{(1 + 0.55\,r^2)^{1.25}}\,\mathrm{sech}(r)$
$g(x, y)$	$0.444\,(1 + 0.04\,r^2)^{0.25}\,\mathrm{sech}(r)$
$h(x, y)$	$0.2\dfrac{0.064\,r^2 + 0.0628r^4}{1 + 4\,r^2 + 0.75\,r^4 + 0.04\,r^6}$
$\hat{g}(x, y)$	$(1 + 3r^2)^{0.25}[0.167\,\mathrm{sech}(r) - 0.295\,\mathrm{sech}(1.3r) - 0.39\,\mathrm{sech}(2.2r)]$
$\hat{h}(x, y)$	$\dfrac{-0.33r^2}{(1 + r^2)^2}[0.87\,\mathrm{sech}(1.89r) + 0.02r^2\,\mathrm{sech}(1.49r) - 0.00078]$

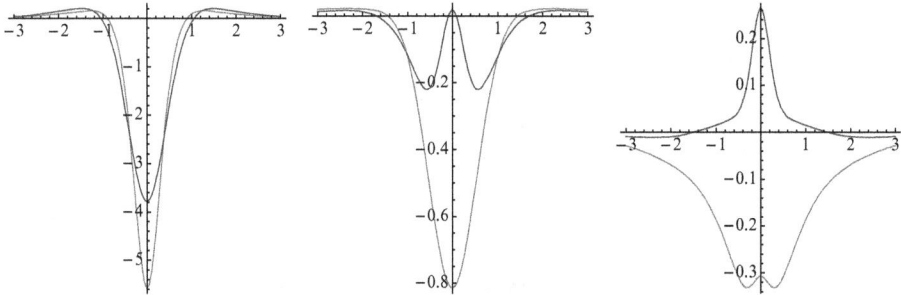

Figure 1.20. Cross sections at $y = 0$ of the Laplacians $\Delta\,F(r)$ (left panel), $\Delta G(r)$ (middle panel), and $\Delta\,H(r)$ (right panel) for the two selected values of σ: 0.95 (red); 3/16 (blue).

for the contour lines. In addition, when β_1 is increased from 0 to 3, the contraction becomes even more pronounced. This is similar to the case of BE with quadratic nonlinearity [24]; in the present case, no depressions are observed before and after the main shape. This means that the nonlinearity selected here is more adequate in a physical sense, and is better suited to the purpose of investigating permanent waves.

Conclusions

In this present paper, we apply to the case of the Boussinesq equation with cubic and quintic nonlinearity a perturbation technique based on the asymptotic expansion for small phase speed, c, and carry out the solution including terms up to $O(c^2)$. Within the adopted asymptotic order, we reduce the original 2D problem to three fourth-order ODEs for functions that depend only on the radial variable. Following our previous work, we construct special approximations on staggered grids which satisfy automatically the behavioral boundary conditions.

Our results confirm that even for this more intricate nonlinearity, the asymptotic decay of the wave profile is algebraic rather than exponential (which is the case with the axisymmetric profile of the standing 2D soliton). This means that the profile of the 2D soliton is not robust: even the presence of a very small phase speed makes the behavior of the shape in the far field qualitatively different.

Similarly to the Boussinesq equation with quadratic nonlinearity (investigated in previous works of Christov), the profile is contracted relatively in the direction of motion and the contraction increases with the increase of the phase speed. The role

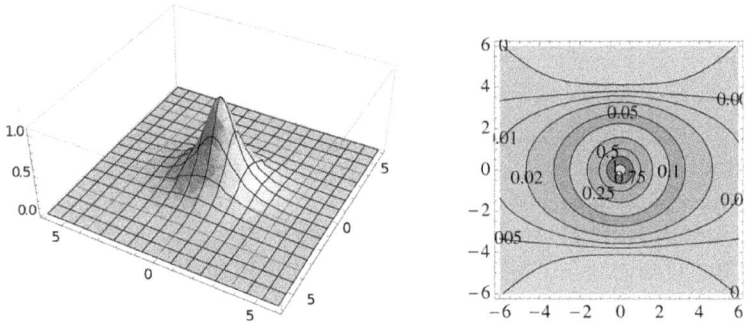

(a) $\sigma = 3/16$, $c = 0.6$, $\beta_1 = 0$.

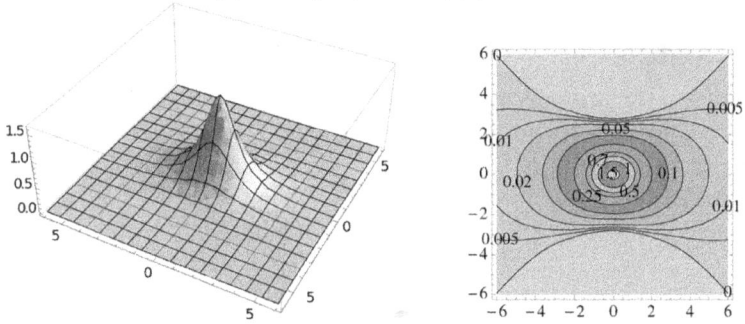

(b) $\sigma = 3/16$, $c = 0.6$, $\beta_1 = 3$.

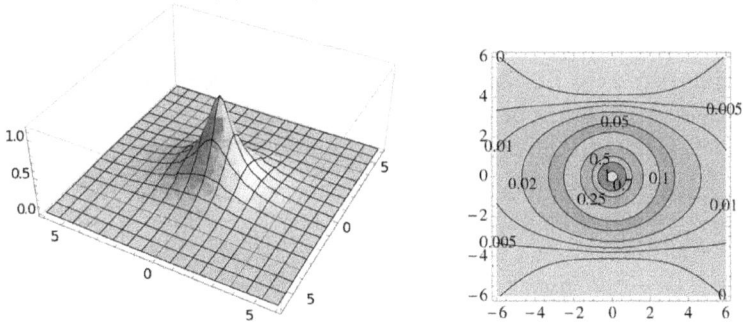

(c) $\sigma = 0.95$, $c = 0.6$, $\beta_1 = 0$.

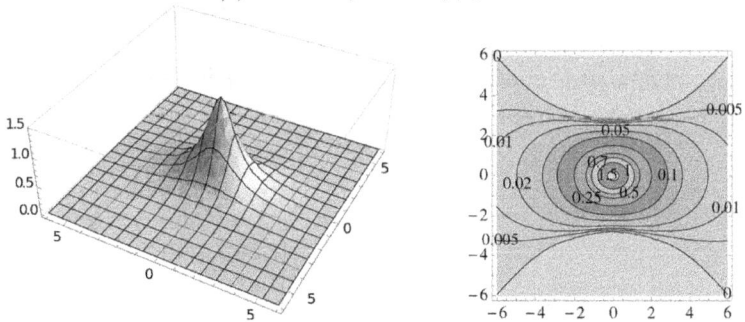

(d) $\sigma = 0.95$, $c = 0.6$, $\beta_1 = 3$.

Figure 1.21. Results for different parameters. (Left) Surface plot. (Right) Contours.

of the dispersion parameter β_1 is shown to enhance the contraction. For larger β_1, the contraction is more pronounced. The most important difference from the case with quadratic nonlinearity is that no depressions (negative height of the profile) are formed in the front and back of the main hum.

The main utility of the solution obtained here is as a precise initial condition for time stepping algorithms that may be used to investigate whether the two dimensional shapes can be permanent, i.e. quasi-particles.

1.5 Boussinesq paradigm equation and the experimental measurement

Let us consider 1D Boussinesq paradigm equation, equation (1.4). The considerations for the DWS model can be applied for this equation also. To this end, we will keep in mind that it has a Hamiltonian structure and can be presented as the following system

$$u_t = q_{xx},$$
$$q_t = u + \alpha u^2 - \frac{\beta}{6}u_{xx} + \frac{\beta}{2}u_{tt}. \tag{1.34}$$

We will use this model in order to simulate a real experiment of observed and measured interacting shallow-water solitary waves. The experiment was conducted in the Nonlinear Waves Lab by Embrie-Riddle Aeronautical University, Daytona Beach, FL. Comparing the profile of the experimental wave in the initial moment $t = 0$ in figure 1.22 with the localized wave, equation (1.35),

$$u = \frac{a}{\cosh^2[b(x - ct)]} \quad \text{with} \quad a = \frac{c^2 - 1}{\alpha}, \quad b = \sqrt{\frac{c^2 - 1}{2\beta\left(c^2 - \frac{1}{3}\right)}}, \tag{1.35}$$

we establish that a best fit is reached for coefficients $\alpha = 0.785$, $\beta = 0.56$ when the phase velocity $c = 1.19$ (dimensionless). The measured velocity of another solitary wave is $c = 1.08$. We simulate head-on collision of two solitary waves that move in the opposite directions (figures 1.23 and 1.24). With insignificant radiation, two

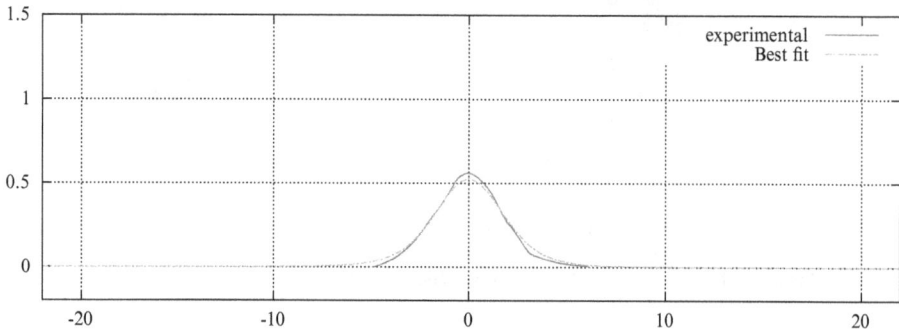

Figure 1.22. Best fit of the experimental measurement for phase velocity $c = 1.19$: $\alpha = 0.785$, $\beta = 0.56$.

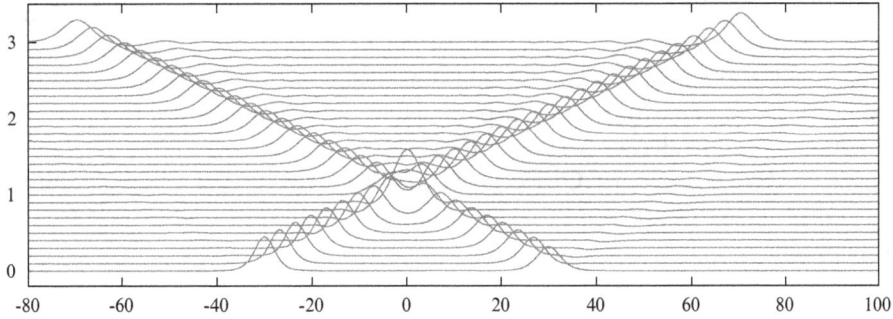

Figure 1.23. Head-on collision of two best-fitted solitons: $c_l = 1.19$, $c_r = -1.08$, $\alpha = 0.785$, $\beta = 0.56$.

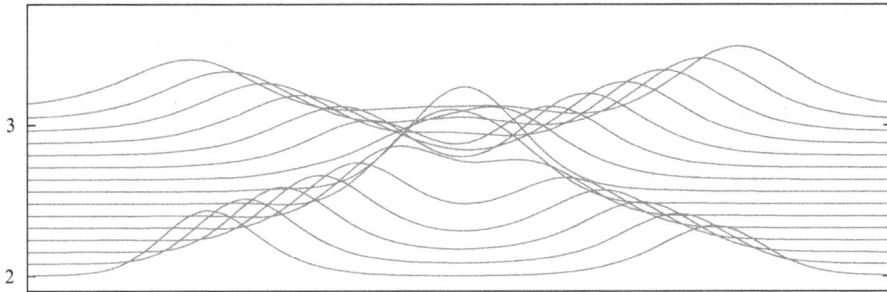

Figure 1.24. Zoom view of the head-on collision from figure 1.23.

solitons interact elastically without a velocity shift. The initial conditions are formulae, equation (1.35). We use the same conservative difference scheme with internal iterations with respect the nonlinear terms as for DWS model.

1.6 Development and realization of efficient numerical methods, algorithms and scientific software for 2D nonsteady Boussinesq paradigm equation. Comparative analysis of the results

Several approaches have been implemented—three finite-difference, one spectral and one pseudospectral. Their numerical realization is based on efficient numerical methods, operator splitting, series expansions and integral transformations.

First approach

The 2D Boussinesq paradigm equation with quadratic nonlinearity

$$u_{tt} = \Delta[u - F(u) + \beta_1 u_{tt} - \beta_2 \Delta u], \qquad F(u) := \alpha u^2,$$

is rendered to a mixed elliptic–hyperbolic system. Trilayer implicit unconditionally stable difference schemes over quasi-uniform spatio-temporal grids of second order are built. Besides, when the explicit discretization of the nonlinear source term does not preserve the energy, a new energy preserving scheme is constructed linearized by internal iterations. The resulting systems of linear equations are solved by an iteration method of the preconditioned conjugate gradient kind. Numerous experiments aimed at the investigation of temporal behavior of 2D soliton structures are conducted. Depending on the given parameters, they demonstrate their instability and transforming to decade waves (self-dispersion) or blow-up [25, 31, 36]. When $F(u) = u^3 - \sigma u^5$, $0 < \sigma < 1$ CQBPE gives decade waves for the initial conditions in [30]. Unlike the quadratic nonlinearity in this case, a blow-up of the solution is not observed [36].

Second approach

Another way to solve 2D BPE consists of implicit difference schemes based on operator splitting via factorization. In this case, the approximate solution is obtained in three steps: in the beginning, two linear algebraic systems with five-diagonal band matrices are solved, each in its coordinate direction, while in the third step, the 2D elliptic problem with discrete Laplacian is solved. The nonlinear part is linearized again by internal iterations. The linear algebraic systems are treated by a modified method of Gauss with pivoting and the method of conjugate gradient for the discrete Laplace equation [35, 40].

Third approach

A new vector scheme (multicomponent ADI scheme) to solve BPE is proposed. In each time layer, one gets a vector of two solutions that simultaneously approximate the exact solution. Some properties of this scheme are investigated—it has a second order of spatial approximation and a first one of temporal approximation. The convergence of the both numerical solutions to the exact one in a uniform norm is proved. For the linearized Boussinesq equation, a discrete identity about the energy of the system is derived. The orders of approximation are confirmed numerically [39].

Fourth approach

To solve the 2D BPE, one can use 2D Fourier transform. The main difficulty as we discussed is how to find the image of the nonlinear term (convolution). The idea of the pseudospectral method consists of using of the backward Fourier transform to get the sought function in the original space, to square it, and to transform the square in the configurational space by forward Fourier transform. In this way, we get the set function in the new time stage including the nonlinear part (fully implicit scheme) [47].

Fifth approach

The Christov–Galerkin spectral method in space $L_2(-\infty, \infty)$ is relatively simpler compared to the pseudospectral because the nonlinear term can be expanded as a combination of basis of functions (known as Christov's functions). In [11], the dynamics of soliton waves are studied, which are a solution of a Boussinesq-like equation with a linear restoring force

$$u_{tt} = (\sigma u - u^2 - \gamma_2 U_{xx})_{xx} - \delta u, \qquad \delta > 0.$$

Here, σu_{xx} is the tension acting along the axis of the rod, $\sigma > 0$—axial extension, $c = \sqrt{\sigma}$—characteristic speed of the transverse (shear) waves; $\sigma < 0$—axial compression, Euler buckling, $\gamma_2 u_{xxxx}$—stiffness term, δu—linear restoring term. This equation describes the bending deformations of the elastic rod. Though the equation is of fourth order, it can describe some properties of the solutions of the sixth-order generalized Boussinesq equation.

Comparison of the different approaches

There is a good comparison between the results obtained by first and second approaches. In order to fix the influence of the given boundary conditions, asymptotic boundary conditions are used as well. When the computational domain is large enough, they (plane or asymptotic) do not deform the solution.

The numerical results with the third approach are compared with the results obtained by the first approach by using a conservative scheme. A good qualitative comparison between the numerical solutions is determined.

References

[1] Ablowitz M J and Segur H 1981 *Solitons and the Inverse Scattering Transform* (Philadelphia, PA: SIAM)

[2] Benjamin T B, Bona J L and Mahony J J 1972 Model equations for long waves in nonlinear dispersive systems *Phil. Trans. R. Soc.* A **272** 47–78

[3] Bona J L, McKinney W R and Restrepo J M 2000 Stable and unstable solitary-wave solutions of the generalized regularized long-wave equation *J. Nonlinear Sci.* **10** 603–38

[4] Bona J L and Sachs R L 1988 Global existence of smooth solutions and stability of solitary waves for a generalized Boussinesq equation *Commun. Math. Phys.* **118** 15–29

[5] Bona J L and Soyeur A 1994 On the stability of solitary-wave solutions of model equations for long waves *J. Nonlinear Sci.* **4** 449–70

[6] Boussinesq J V 1871 Théorie de l'intumescence liquide appelée onde solitaire ou de translation, se propageant dans un canal rectangulaire *C. R. Hebd. Seances Acad. Sci.* **72** 755–9

[7] Boussinesq J V 1871 Théorie générale des mouvements qui sont propagés dans un canal rectangulaire horizontal *C. R. Hebd. Seances Acad. Sci.* **73** 256–60

[8] Boussinesq J V 1872 Théorie des ondes et des remous qui se propagent le long d'un canal rectangulaire horizontal, en communiquant au liquide contenu dans ce canal des vitesses sensiblement pareilles de la surface au fond *J. Math. Pures Appl* **17** 55–108

[9] Boyd J P 2001 *Chebishev and Fourier Spectral Methods* 2nd edn (New York: Dover)

[10] Chertock A, Christov C I and Kurganov A 2011 Central-upwind schemes for the Boussinesq paradigm equation *Computational Science & High Performance Computing IV, NNFM* eds E Krause *et al* pp 267–81

[11] Christou M 2011 *Soliton Interactions on a Boussinesq Type Equation with a Linear Restoring Force* AIP CP1404 eds M Todorov and C I Christov (Melville, NY: American Institute of Physics) pp 41–8

[12] Christou M A and Christov C I 2007 Fourier–Galerkin method for 2D solitons of Boussinesq equation *Math. Comput. Simul.* **74** 82–92

[13] Christou M A and Christov C I 2009 Galerkin spectral method for the 2D solitary waves of Boussinesq paradigm equation *AMiTaNS'09* AIP CP1186 eds M D Todorov and C I Christov (Melville, NY: American Institute of Physics) pp 217–25

[14] Christov C I 1982 A complete orthonormal sequence of functions in $L_2(-\infty,\infty)$ space *SIAM J. Appl. Math.* **42** 1337–44

[15] Christov C I 1995 Fourier-Galerkin algorithm for 2D localized solutions *Annu. Univ. Sofia Livre 2. Math. Appl. Inf* **89** 169–79

[16] Christov C I 1990 Flows with coherent structures: application of random point functions and stochastic functional series *Continuum Models and Discrete Systems* vol 1 ed G A Maugin (London: Longman) pp 232–53

[17] Christov C I 1994 Gaussian elimination with pivoting for multidiagonal systems *Internal Report 4* (Reading: University of Reading)

[18] Christov C I 1995 Conservative difference scheme for Boussinesq model of surfacewaves *Proc. ICFD V* eds W K Morton and M J Baines (Oxford: Oxford University Press) pp 343–9

[19] Christov C I 1995 Numerical investigation of the long-time evolution and interaction of localized waves *Fluid Physics, Proc. Summer Schools* eds M G Velarde and C I Christov (Singapore: World Scientific) pp 403–22

[20] Christov C I 2001 An energy-consistent dispersive shallow-water model *Wave Motion* **4** 161–74

[21] Christov C I 2001 Solitary waves with Galilean invariance in dispersive shallow-water flows *Differential Equations and Nonlinear Mechanics* ed K Vajravelu (Dordrecht: Kluwer) pp 49–68

[22] Christov C I 2012 Numerical implementation of the asymptotic boundary conditions for steadily propagating 2D solitons of Boussinesq type equations *Math. Comput. Simul* **82** 1079–92

[23] Christov C I and Bekyarov K L 1990 A Fourier-series method for solving soliton problems *SIAM J. Sci. Stat. Comput.* **11** 631–47

[24] Christov C I and Choudhury J 2011 Perturbation solution for the 2D Boussinesq equation *Mech. Res. Commun.* **38** 274–81

[25] Christov C I, Kolkovska N and Vasileva D 2011 On the numerical simulation of unsteady solutions for the 2D Boussinesq paradigm equation *NMA'2010, LNCS* vol 6046 eds I Dimov et al pp 386–94

[26] Christov C I and Maugin G A 1995 Numerics of some generalized models of lattice dynamics *Nonlinear Waves in Solids (ASME Book N0 AMR137)* eds J Wegner and R Norwood (New York: ASME) pp 374–9

[27] Christov C I and Maugin G A 1995 An implicit difference scheme for the long-time evolution of localized solutions of a generalized Boussinesq system *J. Comput. Phys.* **116** 39–51

[28] Christov C I, Maugin G A and Porubov A 2007 On Boussinesq's paradigm on nonlinear wave propagation *C. R. Mecanique* **335** 521–35

[29] Christov C I and Todorov M D 2013 Investigation of the long-time evolution of localized solutions of a dispersive wave system *Discrete Cont. Dyn. Syst. Suppl.* **2013** 139–48

[30] Christov C I, Todorov M D and Christou M A 2011 Perturbation solution for the 2D shallow-water waves in *AMiTaNS'11*, AIP CP1404 eds M D Todorov and C I Christov (Melville, NY: American Institute of Physics) pp 53–60

[31] Christov C I and Vasileva D 2011 On the numerical investigation of unsteady solutions for the 2D Boussinesq paradigm equation in a moving frame coordinate system BGSIAM'11 *Proc. 6th Annual Meeting of the Bulgarian Section of SIAM December 21-23, 2011* (Sofia: Demetra Ltd.) pp 103–8

[32] Christov C I and Velarde M G 1994 Inelastic interaction of Boussinesq solitons *Int. J. Bifurc Chaos* **4** 1095–112

[33] Chung K C, Evans G A and Webster J R 2000 A method to generate generalized quadrature rules for oscillatory integrals *Appl. Numer. Math.* **34** 85–93

[34] Cole J D 1968 *Perturbation Methods in Applied Mathematics* (Waltham: Blaisdell Pub. Co)

[35] Dimova M and Kolkovska N 2012 Comparison of some finite difference schemes for Boussinesq paradigm equation *Lect. Notes Comput. Sci* **7125** 215–20

[36] Dimova M and Vasileva D 2013 Comparison of two numerical approaches to Boussinesq paradigm equation *NAA'12, LNCS* vol 8236 eds I Dimov and I Farago pp 255–62

[37] Filon L N G 1928–29 On a quadrature formula for trigonometric integrals *Proc. R. Soc. Edin.* **49** 38

[38] Kalitkin N N 1978 *Numerical Methods* (Moscow: Nauka) p 104 (in Russian)

[39] Kolkovska N and Angelow K 2013 A multicomponent alternating direction method for numerical solution of Boussinesq paradigm equation in *NAA'12, LNCS* vol 8236 eds I Dimov and I Farago pp 371–8

[40] Kolkovska N and Dimova M 2012 A new conservative finite difference scheme for Boussinesq paradigm equation *Cent. Eur. J. Math* **10** 1159–71

[41] Kurganov A and Tadmor E 2002 Solution of two-dimensional Riemann problems for gas dynamics without Riemann problem solvers *Numer. Meth. Part. Differ. Equ.* **18** 584–608

[42] Kutev N, Kolkovska N, Dimova M and Christov C I 2011 Theoretical and numerical aspects for global existence and blow up for the solutions to Boussinesq paradigm equation *AMiTaNS'11*, AIP CP1404 eds M D Todorov and C I Christov (Melville, NY: American Institute of Physics) pp 68–76

[43] Maugin G A and Cadet S 1991 Existence of solitary waves in martensitic alloys *Int. J. Eng. Sci.* **29** 243–58

[44] Newell A C 1985 *Solitons in Mathematics and Physics* (Philadelphia, PA: SIAM)

[45] Porubov A V, Maugin G A and Mateev V V 2004 Localization of two-dimensional non-linear strain waves in a plate *Int. J. Nonlinear Mech.* **39** 1359–70

[46] Porubov A V, Pastrone F and Maugin G A 2004 Selection of two-dimensional nonlinear strain waves in micro-structured media *C. R. Mecanique* **332** 513–8

[47] Todorov M D and Christov C I 2012 Numerical implementation of Fourier-transform method for generalized wave equations BGSIAM'11 *Proc. 6th Annual Meeting of the Bulgarian Section of SIAM December 21–22, 2011* (Sofia: Demetra, Ltd.) pp 97–102

[48] Turitzyn S K 1993 Nonstable solitons and sharp criteria for wave collapse *Phys. Rev. E* **47** R13–6

[49] Turitsyn S K 1993 Blow-up in the Boussinesq equation *Phys. Rev. E* **47** R769–99

[50] Liu Yue 1993 Instability of solitary waves for generalized Boussinesq equations *J. Dyn. Differ. Equ.* **5** 537–58

[51] Zabusky N J and Kruskal M D 1965 Interaction of "solitons" in a collisionless plasma and the recurrence of initial states *Phys. Rev. Lett.* **15** 240–3

IOP Concise Physics

Nonlinear Waves

Theory, computer simulation, experiment

M D Todorov

Chapter 2

Systems of coupled nonlinear Schrödinger equations. Vector Schrödinger equation

The system of coupled nonlinear Schrödinger equations (CNLSEs) also termed as the vector Schrödinger equation is a soliton supporting dynamical system. For the first time, it is considered as a model of light propagation in Kerr isotropic media (see, for example, [25]). Along with that, the phenomenology of the equation opens up the prospect of investigating the quasi-particle behavior of the solitons. The latter corresponds to the understanding and explanation of the fundamental processes and phenomena related to the dynamics of nonlinear waves. The initial polarization of the solutions of vector Schrödinger equation and its evolution evolves from the vector nature of the model. The polarization causes many experimental observations and effects [24]. The concept of quasi-particles makes CNLSE a proper model for testing approximate approaches like the variational approximation, coarse-grain descriptions, etc. The existence of exact (analytical) solutions usually is rendered to the more simple models, while for the vector Schrödinger equation such solutions are not known. This determines the role of the numerical schemes and approaches. The first implicit difference Crank-Nicolson scheme is proposed for the scalar Schrödinger equation [97]. This scheme is extended and adapted for the vector Schrödinger equation by using so-called 'internal iteration' [23]. The realization of the 'internal iteration' results in two advantages: a fully implicit difference scheme for the whole operator and an implementation of the conservation laws on a discrete level within the round-off error.

Using the fully implicit conservative Crank–Nicolson scheme in complex arithmetics for the linear part of the system of CNLSE, internal iteration for the nonlinear part studies the propagation and interaction (head-on collision and taking over) dynamics of solitary waves (quasi-particles). An overall investigation of the influence of the initial parameters: phase difference, carrier frequency, phase

doi:10.1088/978-1-64327-047-0ch2

velocity, polarization, on the solutions is carried out. Considering the given two-component soliton envelope with arbitrary (elliptic) initial polarization, we derive a nonlinear conjugate bifurcation system of differential equations. The effect of the coefficients of nonlinear and linear coupling on the interaction is investigated. Various localized solutions are obtained and interpreted in an appropriate way. We establish a polarization discontinuity ('shock') of the individual polarization angles on the place of the interaction. As a result, the full polarization can be changed qualitatively, however, it remains unchanged quantitatively within the round error. We confirm numerically that the scheme is conservative with respect the total mass, pseudomomentum, and energy. Yet, we find out that the system possesses one more conservation law—the total net polarization is also constant. When the linear coupling is nontrivial, a rotational polarization and breathing solitons are present, even without any interaction. Though the total mass is constant, the components of the solitary wave exchange masses in between, i.e. the individual masses are also breathers. A similar effect for the individual polarization angles is observed. Their curves breath but the total polarization also breathe due to the phase shift. The latter can be considered as a generalization of the net conservation law about the total polarization. All the contributions concern the system of CNLSE and its reductions and expansions. They can be systematized as follows

- Posing of Cauchy problem;
- Soliton solutions;
- Modeling of an initial polarization vector and its effect on the dynamics. Elastic and nonelastic interactions;
- Integrability and conservation;
- Physical and mechanical interpretation of the results.

Following the above outline, we build a conservative, with respect the mass, pseudomomentum, and energy implicit scheme [99]. Being of a Crank–Nicolson kind, it is generalized and computed in complex arithmetic. First, we modify the earlier developed solver for inversion of multidiagonal band matrices [21] for complex-valued matrices. Applied for a numerical implementation of the two-component system of CNLSE (vector Schrödinger equation), it accelerates the needed computation time about fourfold, compared to real-valued arithmetic with four partial differential equations (PDEs). Besides the polarization 'shock', the magnitude of nonlinear coupling can cause newly born high energetic solitons during the collision, while the real part of linear coupling is responsible for the self-focusing/self-dispersion of the soliton envelopes.

2.1 Conservative scheme in complex arithmetic for vector nonlinear Schrödinger equations

Preliminaries

The investigation of soliton supporting systems is of great importance both for the applications and for the fundamental understanding of the phenomena associated with the propagation of solitons. Recently, elaborate models, such as CNLSE, appeared in the literature. They involve more parameters and possess richer phenomenology, but, as a rule, are not fully integrable and require numerical approaches. The non-fully-integrable models possess as a rule three conservation laws: for '(wave) mass', (wave) momentum, and energy, and these have to be faithfully represented by the numerical scheme.

The concept of the internal iterations first applied to CNLSE in [23] was extended in [92] in order to ensure the implementation of the conservation laws on a difference level within the round-off error of the calculations. The CNLSE is also investigated numerically in [58]. Here, we generally follow the works [23, 92] but focus on a new complex-variable implementation of the conservative scheme. This allows us to invert (albeit complex-valued) five-diagonal matrices while the real-valued scheme requires the inversion of nine-diagonal matrices [92]. As we emphasized, we generalize the computer code for Gaussian elimination with pivoting developed earlier for real-valued algebraic systems in [21]. This gives a significant advantage in the efficiency of the algorithm.

The numerical validation of the new code includes comparisons with [23, 92] which show that the complex-numbers implementation of the scheme gives identical results with the real-numbers codes. Several featured examples of interacting solitons in CNLSE are elaborated.

Creating and validating a numerical scheme based on complex-number arithmetic is important for the future application of the conservative-scheme approach in two spatial dimensions. The system for the real and imaginary parts of the wave function has an intricate form that precludes using operator splitting in the real-valued version of the algorithm. Without splitting, the needed computational resources in 2D are enormous, which makes the approach rather unpractical. In this instance, the present approach can be a good basis for development of a 2D numerical scheme based on operator splitting.

Coupled nonlinear Schrödinger equations

In optics, the most popular model is the cubic Schrödinger equation, which describes the single-mode wave propagation in a fiber. It has the form

$$i\psi_t + \beta\psi_{xx} + \alpha|\psi|^2\psi = 0, \tag{2.1}$$

where $i = \sqrt{-1}$ and $\psi(x, t)$ are complex valued wave functions. Depending on the sign of coefficient α, the localized solutions of equation (2.1) are either the hyperbolic secants ('bright solitons') or hyperbolic tangents ('dark solitons'). Since the fibers also allow propagation of multiple orthogonally polarized modes, a multi-component version of equation (2.1) has been actively investigated during the last decade. A general form of the CNLSEs reads

$$i\psi_t + \beta\psi_{xx} + [\alpha_1|\psi|^2 + (\alpha_1 + 2\alpha_2)|\phi|^2]\psi + \gamma\psi + \Gamma\phi = 0,$$
$$i\phi_t + \beta\phi_{xx} + [\alpha_1|\phi|^2 + (\alpha_1 + 2\alpha_2)|\psi|^2]\phi + \gamma\phi + \Gamma\psi = 0, \tag{2.2}$$

where β is the dispersion coefficient and α_1 describes the self-focusing of a signal for pulses in birefringent media. Complex-valued coefficients γ and Γ are responsible for the linear coupling between the two equations. Respectively, α_2 governs the non-linear coupling between the equations. It is interesting to note that when $\alpha_2 = 0$, the nonlinear coupling is not presented, despite the fact that some 'cross' terms proportional to α_1 appear in the equations. In fact, when $\gamma = \Gamma = \alpha_2 = 0$, the solution of the two equations are identical, $\psi \equiv \phi$, and equal to the solution of single NLSE, equation (2.1), with nonlinearity coefficient $\alpha = 2\alpha_1$. The sum $(\alpha_1 + 2\alpha_2)$ is sometimes called 'cross-phase modulation'.

Functions ψ and ϕ have various interpretations in the context of optic pulses, including the amplitudes of x and y polarizations in a birefringent nonlinear planar waveguide, pulsed wave amplitudes of the left and right circular polarizations, etc. The quantity γ is called normalized birefringence, and Γ is the relative propagation constant. The presence of the two new parameters, γ and Γ, in equations (2.2) makes the phenomenology of the system (2.2) much richer. In particular, they allow us to study phenomena such as 'self-dispersion' and 'cross-dispersion', dissipation, etc (see [92] and the literature cite therein).

For Γ, $\gamma = 0$, equation (2.2) is alternatively called the Gross–Pitaevskii equation or an equation of Manakov-type. It was solved analytically for the case $\alpha_2 = 0$, $\beta = \frac{1}{2}$ by Manakov [72] via an inverse scattering transform who generalized an earlier result by Zakharov and Shabat [114, 115] for the scalar cubic NLSE (i.e. equation (2.2)-ψ with $\phi(x, t) = 0$). In a birefringent medium, $\alpha_1 + 2\alpha_2$ distinguishes the description of elliptic, circular and linear polarizations. In this more interesting case, integrability is lost, and numerical methods have to be used to study the evolution of the system.

Let us define energy, E, and mass, M, of the wave system as

$$
\begin{aligned}
E = \int_{-\infty}^{\infty} &\left[-\beta(|\psi_x|^2 + |\phi_x|^2) + \frac{\alpha_1}{2}(|\psi|^4 + |\phi|^4) + (\alpha_1 + 2\alpha_2)(|\phi|^2|\psi|^2) \right. \\
&\left. + \gamma(|\psi|^2 + |\phi|^2) + 2\Gamma\Re(\bar{\phi}\psi) \right] dx, \qquad M = \int_{-\infty}^{\infty} (|\psi|^2 + |\phi|^2) dx,
\end{aligned}
\tag{2.3}
$$

respectively. Then, it is readily proved that these quantities are conserved on the solutions of equation (2.2), namely

$$
\frac{dM}{dt} = 0, \qquad \frac{dE}{dt} = 0.
\tag{2.4}
$$

If one is to construct a numerical algorithm, the above conservation laws have to be embodied in the scheme in order to faithfully represent the physics of the problem.

Conservative difference scheme

Consider a uniform mesh in the interval $[-L_1, L_2]$,

$$
x_i = (i - 1)h, \quad h = (L_1 + L_2)/(N - 1) \quad \text{and} \quad t^n = n\tau,
$$

where N is the total number of grid points in the interval and τ is the time increment. Respectively, ψ_i^n and ϕ_i^n denote the value of the ψ and ϕ at the ith spatial point and time stage t^n. Clearly, $n = 0$ refers to the initial conditions.

Our purpose is to create a difference scheme that is not only convergent (consistent and stable), but also reflects the conservation laws equation (2.43). A scheme that satisfies them reads

$$
i\frac{\psi_i^{n+1} - \psi_i^n}{\tau} = \frac{\beta}{2h^2}\left(\psi_{i-1}^{n+1} - 2\psi_i^{n+1} + \psi_{i+1}^{n+1} + \psi_{i-1}^n - 2\psi_i^n + \psi_{i+1}^n\right),
$$
$$
+ \frac{\psi_i^{n+1} + \psi_i^n}{4}\left[\alpha_1\left(|\psi_i^{n+1}|^2 + |\psi_i^n|^2\right) + (\alpha_1 + 2\alpha_2)\left(|\phi_i^{n+1}|^2 + |\phi_i^n|^2\right)\right] \quad (2.5)
$$
$$
+ \frac{\gamma}{2}(\psi_i^n + \psi_i^{n+1}) + \frac{\Gamma}{2}(\phi_i^n + \phi_i^{n+1})
$$

$$
i\frac{\phi_i^{n+1} - \phi_i^n}{\tau} = \frac{\beta}{2h^2}\left(\phi_{i-1}^{n+1} - 2\phi_i^{n+1} + \phi_{i+1}^{n+1} + \phi_{i-1}^n - 2\phi_i^n + \phi_{i+1}^n\right)
$$
$$
+ \frac{\phi_i^{n+1} + \phi_i^n}{4}\left[\alpha_1\left(|\phi_i^{n+1}|^2 + |\phi_i^n|^2\right) + (\alpha_1 + 2\alpha_2)\left(|\psi_i^{n+1}|^2 + |\psi_i^n|^2\right)\right] \quad (2.6)
$$
$$
+ \frac{\gamma}{2}(\phi_i^n + \phi_i^{n+1}) + \frac{\Gamma}{2}(\psi_i^n + \psi_i^{n+1}).
$$

We prove that the scheme in question conserves the discrete analogs of mass and energy, from (2.43). Namely, for all $n \geqslant 0$, we have

$$
M^n = \sum_{i=2}^{N-1}\left(|\psi_i^n|^2 + |\phi_i^n|^2\right) = \text{const},
$$

$$
E^n = \sum_{i=2}^{N-1}\frac{-\beta}{2h^2}\left(|\psi_{i+1}^n - \psi_i^n|^2 + |\phi_{i+1}^n - \phi_i^n|^2\right) + \frac{\alpha_1}{4}\left(|\psi_i^n|^4 + |\phi_i^n|^4\right) \quad (2.7)
$$

$$
+ \frac{\alpha_1 + 2\alpha_2}{2}\left(|\psi_i^n|^2|\phi_i^n|^2\right) + \frac{\gamma_r^\phi}{2}|\phi_i^n|^2 + \frac{\gamma_r^\psi}{2}|\psi_i^n|^2 + \Gamma\Re[\bar{\phi}_i^n\psi_i^n] = \text{const}.
$$

The scheme equations (2.5) and (2.6) cannot be implemented directly, because it is nonlinear with respect to the variables ψ_i^{n+1} and ϕ_i^{n+1}. Following the idea of [23], we introduce internal iterations, namely

$$
i\frac{\psi_i^{n+1,\,k+1} - \psi_i^n}{\tau} = \frac{\beta}{2h^2}\left(\psi_{i-1}^{n+1,\,k+1} - 2\psi_i^{n+1,\,k+1} + \psi_{i+1}^{n+1,\,k+1} + \psi_{i-1}^n - 2\psi_i^n + \psi_{i+1}^n\right)
$$
$$
+ \frac{\psi_i^{n+1,\,k} + \psi_i^n}{4}\left[\alpha_1\left(|\psi_i^{n+1,\,k+1}||\psi_i^{n+1,\,k}| + |\psi_i^n|^2\right)\right.
$$
$$
\left. + (\alpha_1 + 2\alpha_2)\left(|\phi_i^{n+1,\,k+1}||\phi_i^{n+1,\,k}| + |\phi_i^n|^2\right)\right] \quad (2.8)
$$
$$
+ \frac{\gamma}{2}(\psi_i^n + \psi_i^{n+1,\,k+1}) + \frac{\Gamma}{2}(\phi_i^n + \phi_i^{n+1,\,k+1})
$$

$$i\frac{\phi_i^{n+1,\,k+1} - \phi_i^n}{\tau} = \frac{\beta}{2h^2}\Big(\phi_{i-1}^{n+1,\,k+1} - 2\phi_i^{n+1,\,k+1} + \phi_{i+1}^{n+1,\,k+1} + \phi_{i-1}^n$$
$$- 2\phi_i^n + \phi_{i+1}^n\Big)$$
$$+ \frac{\phi_i^{n+1,\,k} + \phi_i^n}{4}\Big[\alpha_1\big(|\phi_i^{n+1,\,k+1}||\phi_i^{n+1,\,k}| + |\phi_i^n|^2\big) \qquad (2.9)$$
$$+ (\alpha_1 + 2\alpha_2)\big(|\psi_i^{n+1,\,k+1}||\psi_i^{n+1,\,k}| + |\psi_i^n|^2\big)\Big]$$
$$+ \frac{\gamma}{2}\Big(\phi_i^n + \phi_i^{n+1,\,k+1}\Big) + \frac{\Gamma}{2}\Big(\psi_i^n + \psi_i^{n+1,\,k+1}\Big).$$

Let us note that the internal iteration is a kind of linearization of the original equations and we can consider it as an original implementation of the classical Picard iteration applied for wave and dispersive equations. Now for the current iteration of the unknown functions (superscripts $n + 1$, $k + 1$), we have an implicitly coupled system of two tridiagonal systems with complex coefficients. The coupling here is essential. Without it, it is hard to secure absolute stability of the scheme.

We conduct the internal iterations (repeating the calculations for the same time step $(n + 1)$ with increasing value of the superscript k) until convergence, i.e. when both the following criteria are satisfied

$$\max_i|\psi_i^{n+1,\,k+1} - \psi_i^{n+1,\,k}| \leqslant 10^{-12}\max_i|\psi_i^{n+1,\,k+1}|,$$
$$\max_i|\phi_i^{n+1,\,k+1} - \phi_i^{n+1,\,k}| \leqslant 10^{-12}\max_i|\phi_i^{n+1,\,k+1}|. \qquad (2.10)$$

After the internal iterations converge, one gets the solution of the nonlinear scheme by setting $\psi^{n+1} \equiv \psi^{n+1,k+1}$ and $\phi^{n+1} \equiv \phi^{n+1,k+1}$. We mention here that for physically reasonable time increments τ, the number of internal iterations needed for convergence is four to six, which is a small price to pay to have fully implicit, nonlinear and conservative scheme.

Scheme implementation and validation

As mentioned above, the linearized scheme equations (2.8) and (2.9) are inextricably coupled. In order to be solved, the respective two tridiagonal linear algebraic systems are to be recast as a single system. We use a new vector of unknowns

$$\mu_{2i} = \psi_i^{n+1,\,k+1}, \quad \mu_{2i+1} = \phi_i^{n+1,\,k+1}, \qquad i = 1,\ldots N,$$

which is twice as long as the original vectors, ψ_i and ϕ_i. Respectively, the band of the matrix increases to five.

$$
\begin{pmatrix}
1 & \cdots & & & & & & & & \cdots & 0 \\
0 & 1 & & & & & & & & \cdots & 0 \\
0 & \cdots & & & & & & & & \cdots & 0 \\
0 & \cdots & \dfrac{\beta}{2h^2} & 0 & -\dfrac{\beta}{h^2}+\dfrac{i}{\tau}+\dfrac{\gamma_\psi}{2}+f_l & \dfrac{\Gamma_\phi}{2} & \dfrac{\beta}{2h^2} & 0 & \cdots & 0 \\
0 & \cdots & 0 & \dfrac{\beta}{2h^2} & \dfrac{\Gamma_\psi}{2} & -\dfrac{\beta}{h^2}+\dfrac{i}{\tau}+\dfrac{\gamma_\phi}{2} & 0 & \dfrac{\beta}{2h^2} & \cdots & 0 \\
 & & & & & +f_{l+1} & & & & \\
0 & \cdots & & & & & & & \cdots & 0 \\
0 & \cdots & & & & & & & & 1 & 0 \\
0 & \cdots & & & & & & & & \cdots & 1
\end{pmatrix}
\tag{2.11}
$$

$$
\times
\begin{pmatrix}
\mu_1 \\ \mu_2 \\ \cdots \\ \mu_l \\ \mu_{l+1} \\ \cdots \\ \mu_{2N-1} \\ \mu_{2N}
\end{pmatrix}
=
\begin{pmatrix}
0 \\ 0 \\ \cdots \\
\left(\dfrac{i}{\tau}-f_l-\dfrac{\gamma_\psi}{2}+\dfrac{\beta}{2h^2}\Lambda_{xx}\right)\psi_l^n-\dfrac{\Gamma_\phi}{2}\phi_l^n \\
\left(\dfrac{i}{\tau}-f_{l+1}-\dfrac{\gamma_\phi}{2}+\dfrac{\beta}{2h^2}\Lambda_{xx}\right)\phi_l^n-\dfrac{\Gamma_\psi}{2}\psi_l^n \\
\cdots \\ 0 \\ 0
\end{pmatrix}
$$

where $l = 1, 2, \ldots 2N$ and the first two and the last two lines acknowledge the trivial boundary conditions imposed on the unknown functions.

For the inversion of the five-diagonal $2N \times 2N$ matrix, equation (2.11), we created an algorithm based on Gaussian elimination with pivoting, which is a generalization of a similar algorithm for the real system developed in [21].

As should have been expected, the computational time needed for the scheme with complex arithmetic to complete the calculations for a given set of parameters is four times shorter than the scheme with real arithmetic.

The scheme was thoroughly validated through the standard numerical tests involving halving the spacing and time increment. The results confirm the second order of accuracy in space and time. Another crucial validation is possible, through direct comparison with the results [23, 92]. We did repeat a couple of the more involved cases using both the scheme with real arithmetic and the presented scheme with complex arithmetic. The results with the same scheme parameters turn out to be indistinguishable within the order of round-off error.

Results and discussion

We assume the initial condition to be of the form of a single propagating soliton for each of the functions

$$
\psi(x, t),\ \phi(x, t) = A\,\text{sech}[b(x - X - ct)]\exp\left[i\left(\frac{c}{2\beta}x - X - nt\right)\right].
$$

$$
b^2 = \frac{1}{\beta}\left(n + \gamma - \frac{c^2}{4\beta}\right), \qquad A = b\sqrt{\frac{2\beta}{\alpha_1}}, \qquad u_c = \frac{2n\beta}{c},
\tag{2.12}
$$

and n is the carrier frequency. Here, X is the spatial position where the soliton is situated in the initial moment. It is different for the different functions ψ, ϕ. Note also that when n and γ are selected, then one has to check if $c < 2\sqrt{\beta(n+\gamma)}$. Respectively, b^{-1} is a measure of the support of the localized wave.

First, we begin with investigating the role of the nonlinear coupling parameter α_2. As already mentioned above, $\alpha_2 = 0$ means that there is no interaction between the two orthogonal modes ψ and ϕ, despite the fact that $\alpha_1 \neq 0$ means that terms proportional to $|\psi|^2$ are present in the equation for ϕ, and vice versa. It is important to verify this behavior in the actual numerical solution. Indeed, as shown in figure 2.1, no cross-signals are excited during the interaction for this case. It is interesting to note that the effect of α_2 is not monotone. In figure 2.2, we see that the amplitude of the excited secondary cross-signals for $\alpha_2 = 0.1$ is slightly larger than the respective amplitude in figure 2.3, where $\alpha_2 = 0.25$. Yet, the energy of the excited additional oscillation is larger for larger α_2.

The trend of increasing of the additional oscillations is clearly seen in figure 2.4 for $\alpha_2 = 0.5$. The genesis of a standing soliton in the place of the collision is observed. The further increase of $\alpha_2 = 1$ brings into view a new phenomenon: the excitation of faster solitons of smaller amplitude. In addition, the standing soliton in the point of collision becomes more intensive and its support becomes shorter, commensurate

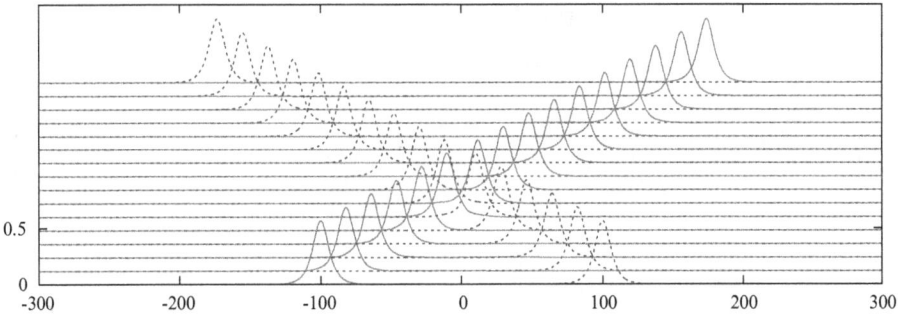

Figure 2.1. Head-on collision for equal carrier frequencies and phase speeds: $n_{\text{left}} = n_{\text{right}} = 0.05$; $c_{\text{left}} = 0.2$, $c_{\text{right}} = -0.2$; $\alpha_1 = 0.25$, $\alpha_2 = 0$, $\beta = 1$, $\Gamma = \gamma = 0$ and number of grid points 3600.

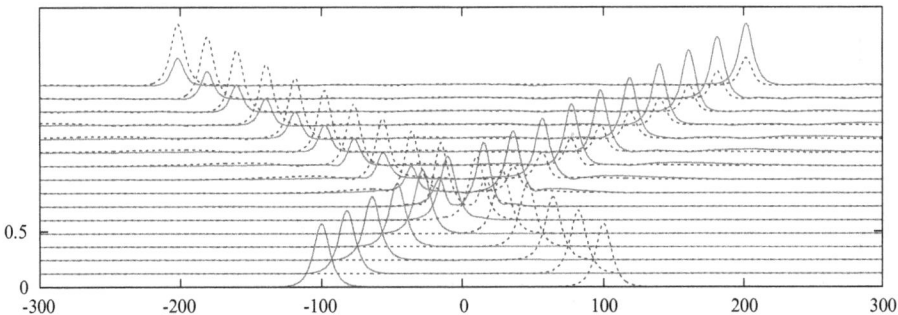

Figure 2.2. The same as in figure 2.1 with $\alpha_2 = 0.1$.

with the support of the original two solitons. The excitation of additional, faster solitons for strong nonlinear interaction $\alpha_2/\alpha_1 = 4$ is a new effect, which is observed for the first time. It was possible because of the strict conservative properties and the high efficiency of the developed numerical scheme.

Another important feature of the collision, for a case when $\alpha_2 > \alpha_1$, is that alongside with the excited new pair of coupled solitons, the main solitons become narrower after the collision. The comparison of figures 2.5 and 2.6 shows that when $\alpha_2/\alpha_1 = 8$, even the secondary solitons become narrower. This effect needs a special investigation and will be given the proper consideration in a separate work.

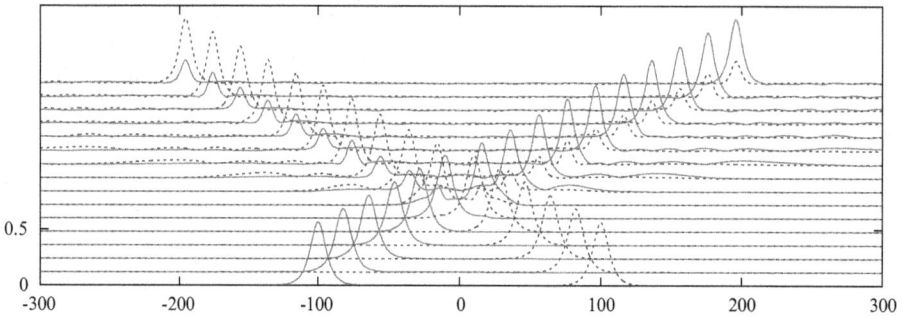

Figure 2.3. The same as in figure 2.1 with $\alpha_2 = 0.25$.

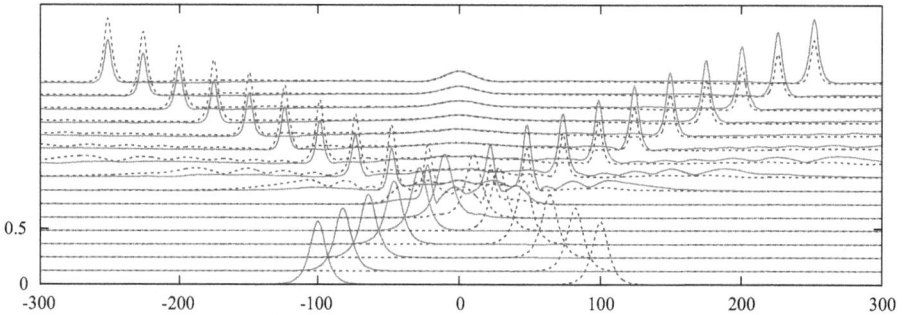

Figure 2.4. The same as in figure 2.1 with $\alpha_2 = 0.5$.

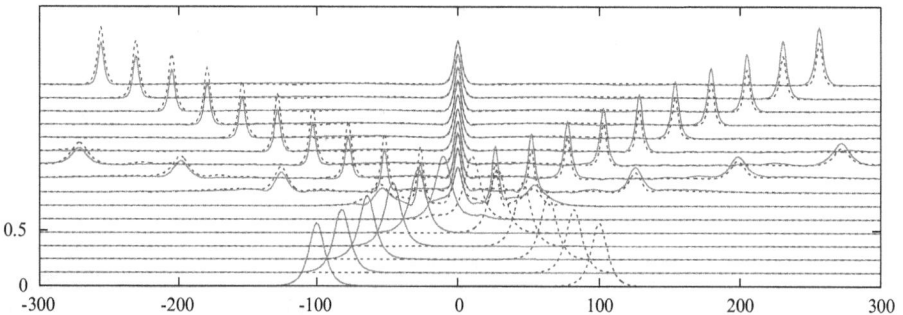

Figure 2.5. The same as in figure 2.1 with $\alpha_2 = 1$.

The non-monotone effect of the nonlinearity is seen in figure 2.7, where the interaction of two non-equal solitons is presented. The smaller (and faster) soliton acquires a larger associated signal than the larger (and slower) soliton. In addition, the amplitudes of these cross-signals are much smaller than in the previous case. Clearly, the amplitudes of the excited cross-signals depend strongly on the total energy of the system.

In the end, we like to elucidate the role of γ when it is strictly real. As mentioned in [92], the imaginary part of γ plays the role of dissipation/energy input according to its sign. The real part of γ does not destroy the conservative nature of the system. The visual comparison of figures 2.3 and 2.8 shows that introducing $\gamma = 0.02$ does not

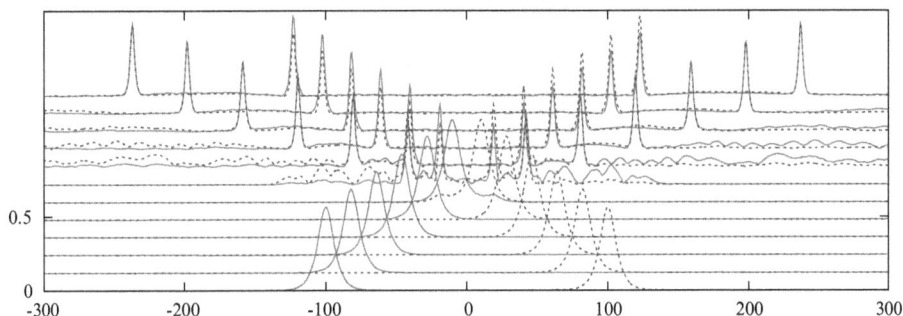

Figure 2.6. The same as in figure 2.1 with $\alpha_2 = 2$.

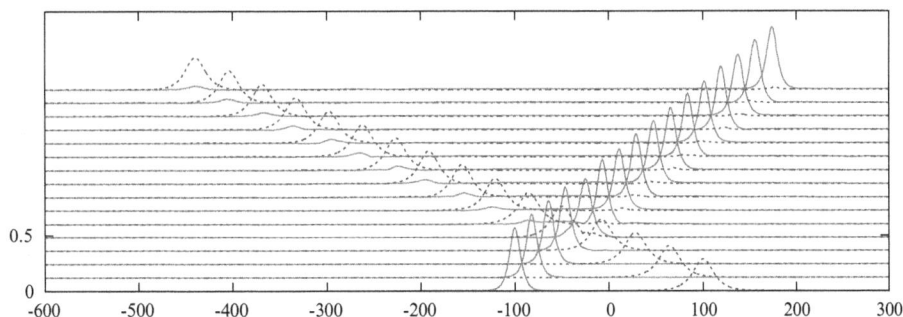

Figure 2.7. Head-on collision for equal carrier frequencies: $n_{\text{left}} = n_{\text{right}} = 0.05$; $c_{\text{left}} = 0.2$, $c_{\text{right}} = -0.4$; $\alpha_1 = \alpha_2 = 0.25$, $\beta = 1$, $\Gamma = \gamma = 0$ and number of grid points 3600.

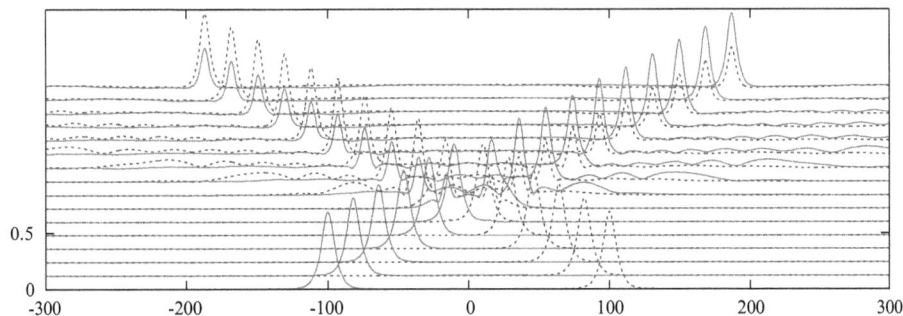

Figure 2.8. The same as in figure 2.3 ($\alpha_2 = 0.25$) but for $\gamma = 0.02$.

qualitatively change the interaction of the solitons, save the fact that the increased self-focusing makes the soliton steeper. For $\gamma < n$, the general behavior is the same as testified by figure 2.9, where the profile is plotted at particular moments of time for several different γ. There are some small differences in the phase shift but they do not warrant a special investigation.

What is more important is that increasing γ to 0.05 (which is in fact the value of n), we get strong amplification of the nonlinear interaction. As shown in figure 2.10, the interaction for $\gamma = 0.05$ and $\alpha_2 = 0.25$ exhibits the birth of the standing soliton, which happens for $\gamma = 0$ only for $\alpha_2 \approx 0.6$–0.7.

Conclusions

We develop a complex-arithmetic implementation of a conservative difference scheme for CNLSEs. To this end, a special solver for Gaussian elimination with pivoting is developed for inverting five-diagonal complex-valued matrices, which is a generalization of the solver created earlier by one of the authors.

The algorithm is validated by the mandatory numerical tests involving doubling the mesh size and halving the time increment, as well as by direct comparison in a couple of cases with the real-arithmetic schemes [23, 92]. The advantages of the new scheme are that the band of the matrix is twice as small and that the overall number of unknowns is also twice as small. Our numerical tests show that, as expected, it performs four times faster than the scheme from [23, 92].

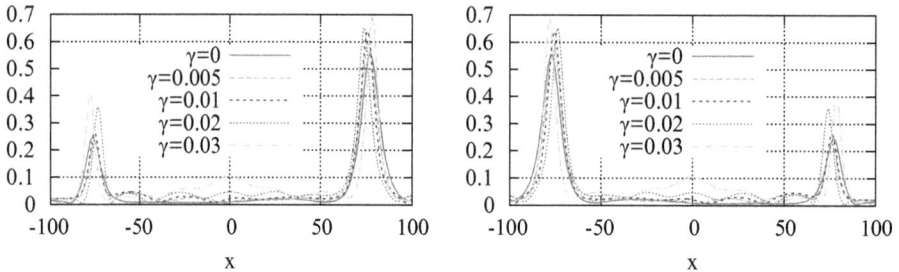

Figure 2.9. Profiles of functions ψ (left) and ϕ (right) at $t = 840$ and different values of γ: $n_{\text{left}} = n_{\text{right}} = 0.05$; $c_{\text{left}} = 0.2$, $c_{\text{right}} = -0.2$, $\alpha_1 = \alpha_2 = 0.25$, $\beta = 1$.

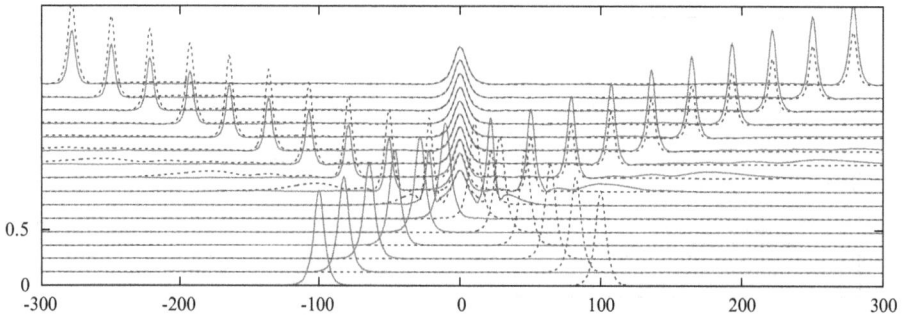

Figure 2.10. The same as in figure 2.3 ($\alpha_2 = 0.25$) but for $\gamma = 0.05$.

The new tool developed here allow us to investigate physically important sets of parameters of the CNLSE under consideration. The nonlinear coupling results in changing the original polarization of the two signals from vertical/horizontal to a generally slanted one. This means that although the initial conditions are nontrivial for only one of the functions in any of the locations, after the interaction both functions acquire nontrivial amplitudes in both locations. First, we unearth the influence of the nonlinear coupling parameter α_2 and show that when it is twice and more times larger than the main nonlinear parameter, α_1 the collision of the two main solitons produces two more solitons that are smaller and faster. For even larger α_2, a standing soliton is born in the exact place of the interaction and it persists without changing shape or dispersing, while the moving solitons continue to the boundaries of the region. The creation of additional quasi-particles appears to be a new effect whose investigation was possible because of the strict conservative properties of the scheme.

Second, we investigate the role of self-focusing (parameter γ). We dwell only on the case when this parameter has a real value, because the imaginary part brings dissipation/energy input and the system ceases to be conservative. For $\gamma < n$ (n is the carrier frequency), the role of the self-focusing is to make the solitons steeper, but the overall qualitative picture of the interaction remains the same. For $\gamma \geqslant n$, the self-focusing also acts to amplify the effect of the nonlinear coupling and additional structures, such as standing solitons, which can appear after the interaction for much smaller α_2 than for the case of $\gamma = 0$.

It is important to stress the point that even new solitons (quasi-particles) are born after the interactions, and the total energy is conserved within the round-off error.

2.2 Finite-difference implementation of conserved properties of the vector nonlinear Schrödinger equation (VNLSE)

We consider the conserved properties of the VNLSE for linearly polarized solitons in the initial configuration. We derive analytic formulae for the mass, pseudomomentum and energy and compare results with the discrete formulae based on a conservative fully implicit finite-difference scheme in complex arithmetic considered in the previous section.

The investigation of soliton supporting systems is of great importance, both for the applications and the fundamental understanding of the phenomena associated with propagation of solitons. Recently, elaborate models, such as the NLSE appeared in the literature (see, for example, [73, 76]). They involve more parameters and have richer phenomenology but, as a rule, are not fully integrable and require numerical approaches. The non-fully-integrable models possess as a rule three conservation laws: for (wave) 'mass', (wave) momentum, and energy and these have to be faithfully represented by the numerical scheme. We follow generally the works [23, 92] but focus on the conservative properties of the conservative scheme. The comparison of both analytic and numerical calculations shows a significant advantage in the efficiency of the finite difference scheme and algorithm.

Vector Schrödinger equation

Here, it is worth mentioning that two main versions of equation (2.1) appear in the literature. In the first one, the sign of the time derivative is positive (as in equation (2.1), which is the most popular version in nonlinear optics), and in the other, the sign is negative. Unlike the parabolic equations, changing the sign does not make the equation incorrect in the sense of Hadamard. Hence, it does not really make any difference which version will be used.

Let us define (pseudo)momentum, P, of the wave system as follows

$$P \stackrel{\text{def}}{=} - \int_{-\infty}^{\infty} \Im(\psi \bar{\psi}_x + \phi \bar{\phi}_x)dx, \tag{2.13}$$

and mass, M, and energy, E, as in (2.3). It is readily proved that the mass, M, and energy, E, are either conserved on the solutions of equation (2.2) (see equation (2.3)), while for pseudomomentum, P, a balance law holds, namely

$$\frac{dP}{dt} = \mathcal{H}\mid_{x=L_2} - \mathcal{H}\mid_{x=-L_1}, \tag{2.14}$$

where $-L_1$ and L_2 are the left end and the right end of the interval under consideration. For asymptotic boundary conditions, the requirement ψ, $\phi = 0$ at infinity entails the requirement that the spatial derivatives also vanish. As a result, the Hamiltonian density vanishes at infinities and the balance law for the pseudo-momentum becomes a conservation law.

We assume that for each of the functions ϕ, ψ, the initial condition is of the form of a single propagating soliton, equation (2.12).

We investigate the evolution of systems of waves which in the initial moment of time are superpositions of solitons of type of equation (2.12) and which evolve according to the system equations (2.2).

To solve this problem numerically, we use the conservative scheme in complex arithmetic described in [99]. If one is to construct a numerical algorithm, the above conservation laws have to be embodied in the scheme in order to faithfully represent the physics of the problem. We use a different number of points in the spatial direction, typically of the order of 8000–20000 points.

The parametric space of the problem is multidimensional, and it is impossible to exhaust the different ranges in a single paper. We focus our attention here on the effect of the nonlinear coupling and set $\Gamma = \gamma = 0$. We also fix $\beta = 1$, because, in fact, the independent variable x can be scaled by β and the latter is not an independent parameter. For definiteness, we fix $\alpha_1 = 1$.

The aim of our work is to better understand the particle-like behavior and properties of the localized waves. We call a localized wave a quasi-particle (QP) if it survives the collision with other QPs (or some other kind of interactions) without losing its identity.

Analytic expressions for conserved properties

Let us assume here that the two components of the vector soliton are moving together without changing their relative position. Then, for both ψ and ϕ

components, we have the same phase speed, c, but the amplitudes, A, size of support, b, and frequencies, n, can be different. A propagating soliton of this type is described by the following formulas

$$\psi = A_\psi \operatorname{sech}[b_\psi(x - ct)]\exp\left[i\left(\frac{c}{2\beta}x - n_\psi t\right)\right],$$

$$\phi = A_\phi \operatorname{sech}[b_\phi(x - ct)]\exp\left[i\left(\frac{c}{2\beta}x - n_\phi t\right)\right]. \tag{2.15}$$

When the compound solution, the vector (ψ, ϕ), propagates steadily, the above parameters are related as in equation (2.12).

For the solution of type equation (2.15), the mass is given by the following formula

$$M = \frac{1}{2\beta}\int_{-\infty}^{\infty}(|\psi|^2 + |\phi|^2)dx = \frac{1}{2\beta}A_\psi^2\int_{-\infty}^{\infty}\operatorname{sech}^2[b_\psi(x - ct)]dx$$

$$+ \frac{1}{2\beta}A_\phi^2\int_{-\infty}^{\infty}\operatorname{sech}^2[b_\phi(x - ct)]dx = \frac{1}{2\beta}\frac{A_\psi^2}{b_\psi}\tanh[b_\psi(x - ct)]\Big|_{-\infty}^{\infty}$$

$$+ \frac{1}{2\beta}\frac{A_\phi^2}{b_\phi}\tanh[b_\phi(x - ct)]\Big|_{-\infty}^{\infty} = \frac{1}{\beta}\left(\frac{A_\psi^2}{b_\psi} + \frac{A_\phi^2}{b_\phi}\right). \tag{2.16}$$

For the wave momentum (pseudomomentum), we get in a similar fashion

$$P = \int_{-\infty}^{\infty}\Im[\psi\bar{\psi}_x + \phi\bar{\phi}_x]dx$$

$$= \int_{-\infty}^{\infty}\left(\frac{A_\psi^2 c}{2\beta}\operatorname{sech}^2[b_\psi(x - ct)] + \frac{A_\phi^2 c}{2\beta}\operatorname{sech}^2[b_\phi(x - ct)]\right)dx \tag{2.17}$$

$$= \frac{1}{\beta}\left(\frac{A_\psi^2}{b_\psi} + \frac{A_\phi^2}{b_\phi}\right)c \equiv Mc.$$

The momentum of the quasi-particle is exactly the product of the mass and the phase speed.

Now, for the energy, we get

$$E = \int_{-\infty}^{\infty}\left[\beta(|\psi_x|^2 + |\phi_x|^2) - \frac{1}{2}\alpha_1(|\psi|^4 + |\phi|^4)\right.$$

$$\left. - (\alpha_1 + 2\alpha_2)(|\phi|^2|\psi|^2) - \gamma(|\psi|^2 + |\phi|^2) - 2\Gamma\Re(\bar{\psi}\psi)\right]dx$$

$$= \frac{2}{3}\beta(A_\psi^2 b_\psi + A_\phi^2 b_\phi) + \frac{c^2}{2\beta}\left(\frac{A_\psi^2}{b_\psi} + \frac{A_\phi^2}{b_\phi}\right) - \frac{2}{3}\alpha_1\left(\frac{A_\psi^4}{b_\psi} + \frac{A_\phi^4}{b_\phi}\right)$$

$$- (\alpha_1 + 2\alpha_2)A_\psi^2 A_\phi^2\int_{-\infty}^{\infty}\operatorname{sech}^2[b_\psi(x - ct)]\operatorname{sech}^2[b_\phi(x - ct)]dx$$

$$- 2\left[\gamma\left(\frac{A_\psi^2}{b_\psi} + \frac{A_\phi^2}{b_\phi}\right) + 2\Gamma\frac{A_\psi^2}{b_\psi}\right].$$

In the above formula, one of the integrals cannot be found analytically directly unless $b_\psi = b_\phi$. Yet, we can have a reasonable approximation after we note that

$$\text{sech}^2[b_\psi(x - ct)]\text{sech}^2[b_\phi(x - ct)]$$

$$= \frac{4}{\left\{\cosh[(b_\psi + b_\phi)(x - ct)] + \cosh[(b_\psi - b_\phi)(x - ct)]\right\}^2}.$$

and that for the cases treated in this work $b_\psi \approx b_\phi$, i.e. $|b_\phi - b_\psi| \ll |b_\phi + b_\psi|$. In such a case, in the region where $\cosh[(b_\psi - b_\phi)(x - ct)]$ changes rapidly, one can assume that $\cosh[(b_\psi - b_\phi)(x - ct)] \sim 1$. Then,

$$\int_{-\infty}^{\infty} \frac{4dx}{\left\{\cosh[(b_\psi + b_\phi)(x - ct)] + \cosh[(b_\psi - b_\phi)(x - ct)]\right\}^2}$$

$$\approx \int_{-\infty}^{\infty} \frac{4dx}{[1 + \cosh(b_\psi + b_\phi)(x - ct)]^2} \qquad (2.18)$$

$$= \int_{-\infty}^{\infty} \frac{dx}{\cosh^4\left[\frac{1}{2}(b_\psi + b_\phi)(x - ct)\right]} = \frac{8}{3}\frac{1}{b_\psi + b_\phi}.$$

The accuracy of this formula can easily be verified for a couple of specific values of b_ψ and b_ϕ for which the original integral can also be found analytically. For specific ratios b_ϕ/b_ψ, both the approximate and exact integrals can be represented as κ/b_ψ, where κ is a different coefficient for the different cases. The results for several different ratios b_ϕ/b_ψ show that up to $b_\phi = 2b_\psi$, the approximation is very reasonable—it does not exceed 3.5%, and hence can be used to get an approximate analytical expression for the integral and compare the energy to the numerically obtained value.

Finally, under the above assumption that the scales of the supports for the two components are not very different, we have the following analytical approximation of the energy

$$E \approx \frac{c^2}{2}\left[\frac{1}{\beta}\left(\frac{A_\psi^2}{b_\psi} + \frac{A_\phi^2}{b_\phi}\right)\right] + \frac{2}{3}\beta(A_\psi^2 b_\psi + A_\phi^2 b_\phi) - \frac{2}{3}\alpha_1\left(\frac{A_\psi^4}{b_\psi} + \frac{A_\phi^4}{b_\phi}\right)$$

$$- \frac{8}{3}(\alpha_1 + 2\alpha_2)\frac{A_\psi^2 A_\phi^2}{b_\psi + b_\phi} - 2\left[\gamma\left(\frac{A_\psi^2}{b_\psi} + \frac{A_\phi^2}{b_\phi}\right) + 2\Gamma\frac{A_\psi^2}{b_\psi}\right]. \qquad (2.19)$$

The term $\frac{c^2}{2}[\frac{1}{\beta}(\frac{A_\psi^2}{b_\psi} + \frac{A_\phi^2}{b_\phi})] \equiv \frac{Mc^2}{2}$ can be called the 'kinetic energy' of the quasi-particle, while the rest of the terms can be called the 'internal energy' of the quasi-particle.

Elastic head-on collision

As an illustration, we computed the solution for $\alpha_2 = 0$ (known as Manakov's solution). As expected, no interaction between the two components of the vector soliton was observed, which confirms that only α_2 governs the nonlinear effects, not the full coefficient $(\alpha_1 + 2\alpha_2)$. As should have been expected, our computations showed that for $\alpha_2 = 0$, there was no interaction between the two orthogonal modes ψ and ϕ, despite the fact that $\alpha_1 \neq 0$ means that terms proportional to $|\psi|^2$ are present in the equation for ϕ, and *vice versa*.

We chose for the phase speeds of solitons $c_1 = 1$ and $c_r = -0.5$, which does not restrict us very much because in absence of linear coupling, $\gamma = \Gamma = 0$, one can change the phase speed, but still obtain the same results provided that α_1 is also changed. The selected values for the phase speeds give for the amplitudes of the initial solitons the following

$$A_1 \equiv A_\psi = \frac{\sqrt{2}}{2} \approx 0.7075, \qquad A_r \equiv A_\phi = \frac{\sqrt{2}}{4} \approx 0.3537. \qquad (2.20)$$

According to the analytical expression, equation (2.16), the masses of the two quasi-particles are $M_1 \equiv M_\psi = 1$ and $M_r \equiv M_\phi = 0.5$. Respectively, the total pseudomomentum is $c_1 M_1 - c_r M_r = 0.75$. Since in the initial moment of time the two QPs are strictly $90°$ polarized, we have only one of the amplitudes A_ψ, A_ϕ not equal to zero. Then, equation (2.19) can be applied to the left and right solitons separately to get

$$E_1^k = \frac{1}{2}c_1^2 M_1 = 0.5, \quad E_1^p = \frac{2}{3}\left(A_1^2 b_1 - \frac{A_1^4}{b_1}\right) = \frac{2}{3}\left(\frac{1}{4} - \frac{1}{2}\right) = -\frac{1}{6} \approx -0.1667$$

$$E_r^k = \frac{1}{2}c_r^2 M_r = \frac{1}{16} = 0.0625, \quad E_r^p = \frac{2}{3}\left(A_r^2 b_r - \frac{A_r^4}{b_r}\right)$$

$$= \frac{2}{3}\left(\frac{1}{32} - \frac{1}{16}\right) = -\frac{1}{48} \approx -0.0208$$

$$E_1 = E_1^k + E_1^p = \frac{1}{3} \approx 0.3333, \qquad E_r = E_r^k + E_r^p = \frac{1}{24} \approx 0.0408,$$

$$E = E_1 + E_r = \frac{3}{8} = 0.375$$

where the superscripts stand for 'kinetic' and 'potential' energies.

Note that the actual values obtained from the initial condition after being discretized on the chosen grid (see the discrete analogues in [92] and [100]), are

$$M_1 = 1.0000000, \quad M_r = 0.50000000, \quad P = 0.74921909, \quad E = 0.37462784.$$

The small deviations for P and E of order of 0.1% are the effect of the truncation error. Since the scheme is conservative, the above values are the ones which are kept constant during the time stepping.

Conclusion

We develop a complex-arithmetic implementation of a conservative difference scheme for vector NLSE.

The new tool developed here allowed us to investigate physically important sets of parameters of the vector NLSE. This means that although the initial conditions are nontrivial for only one of the functions in each of the initial locations, after the interaction both functions can acquire nontrivial amplitudes in both locations.

We consider as an initial profile the superposition of solitons with linear polarizations, one of them having only ψ-component, and the other—only a ϕ-component. Then this initial profile is allowed to evolve according to the VNLSE and the results of the collision depend mostly on nonlinear coupling parameter α_2 (cross-modulation parameter) on the dynamics of quasi-particles [100]. The exact formulae for the masses, pseudomomenta, and energy of the initial solitons in the case of initial linear polarization allow us to check validity and relevance of their discrete analogues and to get significant knowledge about the behavior of the interacted soliton envelopes considered as quasi-particles.

We consider collision dynamics of two soliton envelopes with different initial circular polarizations, i.e. $0° < \theta < 90°$ [101]. Unlike the initial conditions used above, where they are a superposition of two linearly polarized envelopes with angles $\theta_{\text{left}} = 0°$ and $\theta_{\text{right}} = 90°$ [99], here the initial condition is more general. Together with the interactions (head-on collisions) of the solitons, the concept about their behavior as quasi-particles is displayed. When the interaction is nonelastic ($\alpha_2 > 0$), the individual polarization angles are discontinued on the place of the interaction. It turns out that this phenomenon depends essentially on the initial phase difference [99, 100]. The results can reveal the inner mechanism of interaction of the polarized quasi-particles as solutions of the vector Schrödinger equation.

2.3 Collision dynamics of circularly polarized solitons in nonintegrable VNLSE

The system of CNLSE has been first formulated for modeling the propagation of light pulses in optical fibers, and is still one of the main tools in theoretical investigation in fiber optics [9, 52]. Another important role the NLSEs play is as a mathematical object on which new approaches for finding analytical or approximate solutions are tested (see, for example, [72, 114]). Still another aspect of this important model is that, similarly to the models based on sine-Gordon (sG), Korteweg–de Vries (KdV), and Boussinesq equations (to mention a few), the generic class of NLSEs offers the opportunity to investigate the quasi-particle (QP) behavior of solitons. QPs hold the key for the future advances in wave-particle dualism. The NLSE class is perhaps more important for the wave-particle duality than the other famous soliton supporting equations, because its soliton solutions are envelopes (pulses). The carrier frequency of the pulses offers an additional dimension for understanding the analogy between waves and particles.

A system of CNLSEs exhibits a rich phenomenology possessing a plethora of different one-soliton solutions. The essential new feature of CNLSEs in comparison with the single NLSE is the polarization, which is related to the relative amplitudes of the two components. Adding the fact that each component actually is a complex-valued function, one can appreciate the complexity of the possible soliton interactions. Having two components that satisfy two *nonlinearly* coupled equations means that the cross-excitation of the modes (components) during the soliton collision will give rise to radically new physics in comparison with the single NLSE. The CNLSE model is even richer when a linear coupling is considered alongside the main, nonlinear, coupling (see [92] and the literature cited therein).

The simplest possible QPs in CNLSE are those with the so-called 'linear polarization', meaning that each QP has only one non-trivial component. Naturally, if the two initial QPs have the same type of linear polarization (say, in both of them, the same component is set to zero), then the initially zero second component is never excited, and the behavior is essentially the one of a single NLSE. Yet, all it simply takes is to invert one of the linear polarizations of the initial QPs, and the coupled nature and excitability of the system show up.

Another important case of analytical one-soliton solution is presented by the so-called 'circularly polarized' solitons when both modes are present. In the cases of both linear and circular polarization, the carrier frequencies of the two components are identical. Another level of complexity is reached when the carrier frequencies are allowed to be different. In this case, the solitons are called 'elliptically polarized' (see, for example, [95]), and no analytical solution exists even for the one-soliton solution. Then even the initial shape of the QP has to be found numerically (see [74, 95] and the literature cited therein). As demonstrated numerically in [99, 100], in most of these cases, the initially linearly polarized QPs end up as elliptically polarized after the collision. In this sense, all the discussed polarizations are actually limiting cases of the general elliptic polarization.

We set the goal to expand our work on the QPs collision in CNLSE to include the other class of analytical initial conditions: the superposition of circularly polarized one-soliton solutions, and to outline the limits of the analytical two-soliton solution given by Manakov [72] for one limiting case of the cross-modulation parameter.

Apart from some limiting cases, CNLSEs are not fully integrable and require numerical approaches for their treating. As a rule, the model admits two conservation laws: for the 'mass' (better called 'pseudomass' when the QPs are considered) and the energy, and a balance law for the wave momentum (called 'pseudomomentum' for the QPs). In the absence of excitations at the boundary of the domain under consideration, the balance law for the pseudomomentum becomes a conservation law too.

The conservation laws are to be faithfully represented by the respective numerical scheme, which means that it *has* to be nonlinear.

Outline of the model

CNLSE can be written in different equivalent forms. We use the following one (see also [100])

$$i\psi_t + \beta\psi_{xx} + [\alpha_1|\psi|^2 + (\alpha_1 + 2\alpha_2)|\phi|^2]\psi = 0, \qquad (2.21a)$$

$$i\phi_t + \beta\phi_{xx} + [\alpha_1|\phi|^2 + (\alpha_1 + 2\alpha_2)|\psi|^2]\phi = 0. \qquad (2.21b)$$

Here, β is the dispersion parameter, α_1 is called self-focusing, and α_2 is the cross-modulation parameter.

The initial condition for a single QP with circular polarization has the form

$$\psi(x,\, t;\, c,\, n,\, X,\, \theta,\, \vec{\delta}\,) = A\cos\theta\, \mathrm{sech}[b(x - X - ct)]\exp\left\{i\left[\frac{c}{2\beta}x + nt + \delta_\psi\right]\right\}, \quad (2.22a)$$

$$\phi(x,\, t;\, c,\, n,\, X,\, \theta,\, \vec{\delta}\,) = A\sin\theta\, \mathrm{sech}[b(x - X - ct)]\exp\left\{i\left[\frac{c}{2\beta}x + nt + \delta_\phi\right]\right\}, \quad (2.22b)$$

where c and n are the phase speed and the carrier frequency of the QP. The localized solution is identified by the presence of a spatial point, $x = ct - X$ ('center'), which moves with a given phase speed c and has an internal variable n—the carrier frequency. In our notations given in equation (2.1), the parameters are separated from the spatio-temporal independent variables by a semicolon. Respectively, θ is called the 'polarization angle', and $\delta_{\psi,\phi}$ is the phase of each component of the vector solution. In order for a one-soliton solution of the above type to exist, the amplitude, A, and the inverse measure of support, b, have to be related to c and n as follows

$$b^2 = \frac{1}{\beta}\left(n + \frac{c^2}{4}\right), \quad A = b\sqrt{\frac{2\beta}{\alpha_1}}, \quad n > -\frac{c^2}{4}. \quad (2.22c)$$

The meaning of the above restriction is that the carrier frequency cannot be smaller than a specified negative number, for a QP that propagates with a given phase speed c. In particular, if we are concerned with a non-propagating envelope $c = 0$, then its carrier cannot be a standing wave, but only a propagating wave with $n \neq 0$.

It is important to note here that if the polarization angle θ is arbitrary, the circularly polarized QP from equation (2.1) exists only for $\alpha_2 = 0$ (the Manakov case [72]). Yet, for the particular case of $\theta = 45°$, the general elliptic polarization is reduced to a circular polarization with amplitude given by

$$A_{45°} = b\sqrt{\frac{\beta}{\alpha_1 + \alpha_2}}. \quad (2.23)$$

For streamlining the notation, it is convenient to introduce the vector notations $\vec{\chi} = (\psi,\, \phi)^T$ and $\vec{\delta} = (\delta_\psi,\, \delta_\phi)^T$.

The initial condition is constructed as the superposition of two QPs situated at X_l and X_r, and propagating with phase speeds c_l and c_r, i.e.

$$\vec{\chi}(x,\, 0) = \vec{\chi}_l(x,\, 0;\, c_l,\, n_l,\, X_l,\, \theta_l,\, \vec{\delta}_l) + \vec{\chi}_r(x,\, 0;\, c_r,\, n_r,\, X_r,\, \theta_r,\, \vec{\delta}_r). \quad (2.24)$$

Here, $|X_r - X_l|$ has to be large enough so the 'tail' of one QP is fairly well decayed in the place of the other QP.

Difference scheme

We use the implicit scheme, equations (2.5) and (2.6), which is not only convergent (consistent and stable), but also conserves mass and energy

Since the above scheme is of a Crank–Nicolson type for the linear terms (involving the new time stage), we employ internal iterations to achieve implicit approximation of the nonlinear terms, i.e. we use its linearized implementation (see the details of this approach in [23, 92, 99]).

For the parameter range of interest, we conduct numerical experiments to identify the adequate grid values. We find that for carrier frequencies $|n| \leqslant 1$ (when the soliton supports are wider) those values are $\tau = 0.04$ and $h = 0.1$. For carrier frequencies $|n| > 1$, the required spacing was $h \in [0.01, 0.05]$ because the support becomes narrower, which requires a denser grid.

Interaction of circularly polarized solitons

The parametric space of the problem is too large to be explored in full, and we choose $n_{l\psi} = n_{r\psi} = n_{l\phi} = n_{r\phi} = -0.5$, $c_l = -c_r = 1$, $\alpha_1 = 0.75$, and focus on the effects of α_2, θ, and $\overrightarrow{\delta}$.

The Manakov case

The Manakov solution in the form described in equation (2.1) is valid only when θ is a constant that does not depend on x and t. This means that in order to have an analytical two-soliton solution (see [60, 86] for the analytical details), the two initial pulses in equation (2.24) must have the same polarization angles. For this reason, we begin our investigation with two pulses of the same polarization. Clearly, $\theta = 0°$ or $\theta = 90°$ need not be considered here because then one of the components is always equal to zero, and the system reduces to a single NLSE: a case thoroughly investigated in the literature.

Concerning the presentation of the results, we show the interaction itself, as well as snapshots of the initial shapes of the QPs. In the lower small subfigures, we present the real and imaginary parts of the initial QPs in order to give an idea about the phase. The middle panel shows the actual evolution of the profile, and the upper small boxes show again the real and imaginary parts of the QPs, but from the last time stage shown in the middle panel.

First, we examine the case $\theta_l = \theta_r = 45°$, because it has relevance both for the systems with and without cross-modulation. In the succession of drawings in figure 2.11, we present the interaction of two pulses (QPs) of the same initial polarization, but whose phases differ. Shown in figure 2.11(a) is the case when both of QPs have zero phases. The interaction follows exactly the analytical Manakov two-soliton solution.

A note is due here about the initial phases shown in the lower small panels of the figures where the real and imaginary parts are depicted. Although, in some cases, the two QPs have zero initial phases, they appear reflected to each other, because the spatial wave number of the right QP is negative. The analytical pulses exist only for $k_r = \frac{1}{2}c_r < 0$.

The surprise comes in figure 2.11(b) where the interaction of two QPs is presented: the right one having a nonzero phase $\delta_r = 45°$. One can see that after the interaction, the two QPs again become Manakov solitons, but different than the original two that entered the collision. The outgoing QPs have polarizations 33° 48′ and 56° 12′. Something that can be called a 'polarization shock' takes place. Note that all the solutions are perfectly smooth, but because the property called polarization cannot be defined in the cross-section of interaction and for this reason, it appears to be undergoing a shock. There is some indication in the analytical works (see [60]) that this kind of exchange of polarization can take place, but the direct comparison is rather technical and goes beyond the scope of this paper.

Figure 2.11(c) presents a case in which the phases of the original QPs differ by 90°. The 'shock' effect on the polarization is even more pronounced than in figure 2.11(b) (see table 2.1 for the numbers). Here, it is worth mentioning that the moduli of ψ and ϕ from figure 2.11(c) perfectly match each other when rescaled, which means that the resulting solitons have circular polarization (see figure 2.16(a) below).

An important quantitative feature of the result for non-zero phase of one of the QPs is that the sum of the outgoing polarizations is equal to 90°, which is the sum of the two in-going polarizations, each being 45°. In other words, our results show that a kind of 'conservation of polarization' law is obeyed (see table 2.1). In most of the cases presented in this paper, the total polarization is conserved with accuracy better than 4′. The superscripts i and f in the table refer to the initial and final values of the polarization angles.

We would like to add here that we did a numerical experiment with $\delta_l = 0°$ and $\delta_r = 180°$ and it was identical to the case $\delta_l = \delta_r = 0°$. We can say that the most pronounced change of the polarization is observed for the difference between the two phases around 90°. In order to verify that the above described behavior is authentic to the system when $\alpha_2 = 0$, we also consider the case $\theta_l = \theta_r = 20°$ shown in figure 2.12 for $\delta_l = 0°$, $\delta_r = 90°$. One sees that once again the desynchronization of the phases leads to the break up of the Manakov solution (transforming it to a superposition of two Manakov one-soliton solutions).

The conclusion is that although the system is integrable for $\alpha_2 = 0$, the analytical solution of Manakov is not unique. Yet, its relevance is obvious, because after the interaction, the two QPs that leave the site of the collision are once again Manakov solitons but with polarizations that are different from the initial polarization.

To complete the case $\alpha_2 = 0$, we investigate the collision of two QPs whose initial polarizations are not equal. Clearly, one cannot expect in this case to get something like the Manakov two-soliton solution, because of the differences in the polarizations. Figure 2.13 shows the result for two different initial phases. Contrary to the case $\alpha_2 = 0$, a zero phase leads to larger disparity of the outgoing polarizations, while a phase of 90° preserves the polarizations. Clearly, the nonlinearity embodied in the cross-modulation terms can change qualitatively the character of the soliton interactions. The last entries of table 2.1 demonstrate the good conservation of the total polarization in this case.

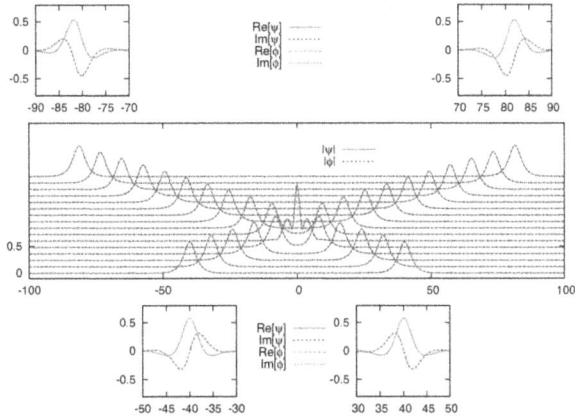

(a) $\delta_l = 0°$, $\delta_r = 0°$

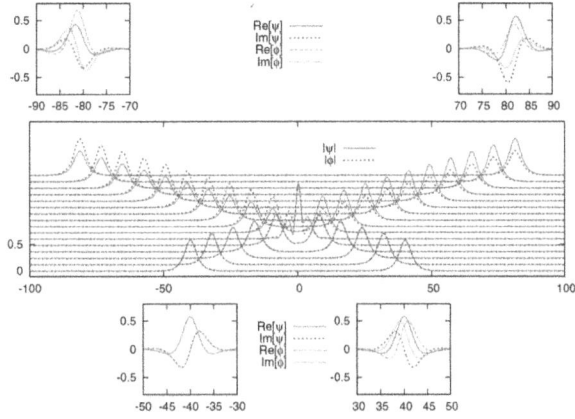

(b) $\delta_l = 0°$, $\delta_r = 45°$

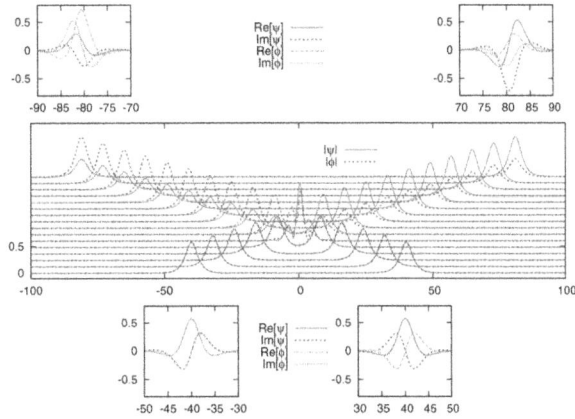

(c) $\delta_l = 0°$, $\delta_r = 90°$

Figure 2.11. Interaction of two QPs with initially equal polarizations $\theta_l = \theta_r = 45°$.

Table 2.1. Conservation of the total polarization.

$\delta_r - \delta_l$	θ_l^i	θ_r^i	$\theta_l^i + \theta_r^i$	θ_l^f	θ_r^f	$\theta_l^f + \theta_r^f$
45°	45°	45°	90°	33° 48′	56° 12′	90°
90°	45°	45°	90°	24° 06′	65° 54′	90°
0°	20°	20°	40°	20° 00′	20° 00′	40°
90°	20°	20°	40°	28° 48′	2° 02′	30° 50′
0°	36°	36°	72°	36° 00′	36° 00′	72°
90°	36°	36°	72°	53° 00′	13° 20′	66° 20′
0°	10°	80°	90°	21° 05′	68° 54′	89° 59′
90°	10°	80°	90°	9° 27′	80° 30′	89° 57′
0°	20°	70°	90°	35° 46′	54° 11′	89° 58′
90°	20°	70°	90°	16° 07′	73° 49′	89° 56′
0°	30°	60°	90°	42° 58′	46° 59′	89° 57′
90°	30°	60°	90°	19° 41′	70° 14′	89° 55′

Case of nontrivial cross-modulation $\alpha_2 \neq 0$

As we have shown in the precedence, an analytical solution with circular polarization is possible even for non-trivial values of cross-modulation parameter, $\alpha_2 \neq 0$, but only for one particular polarization $\theta = 45°$. In the previous subsection, we have found that the phases of the components play an essential role when the system chooses one or another of polarization of the one-soliton solutions that break away from the site of collision. It seems important to understand how the circularly polarized initial QPs evolve in the case of nontrivial cross-modulation parameter. We have computed the collision of two initial QPs with circular polarization $\theta = 45°$ for different α_2 and we have discovered that the interaction obeys the two-soliton solution with polarization $\theta = 45°$ for all of the values of cross-modulation that we have considered. In figure 2.14, we present the case $\alpha_2 = 0.5$ as a featured example of the above statement.

The result shown in the figure is qualitatively similar to figure 2.11(a), save for the different amplitudes of the QPs stemming from equation (2.22c), rather than equation (2.22b). This means that the circularly polarized two-soliton solutions can be accessed via direct numerical simulation even when $\alpha_2 \neq 0$, which is an important new information about the CNLSE systems.

Now, the last question that remains to be answered is what will happen when there is a difference between the initial phases of the QPs. To answer this question, we perform a similar experiment with different phases for $\alpha_2 = 0.25$. Figure 2.15 shows the result which appears similar to the result with $\alpha_2 = 0$ in figure 2.11(c).

When $\alpha_2 \neq 0$, it is not clear in advance that the resulting soliton has circular polarization. In order to judge what is the polarization of the QPs on the final stage of figure 2.15, we show the comparison of the scaled profiles of ψ and ϕ in figure 2.16(b). These profiles do not coincide, which means that they are not *sech*-es, and the polarization is not circular, but is generally elliptic.

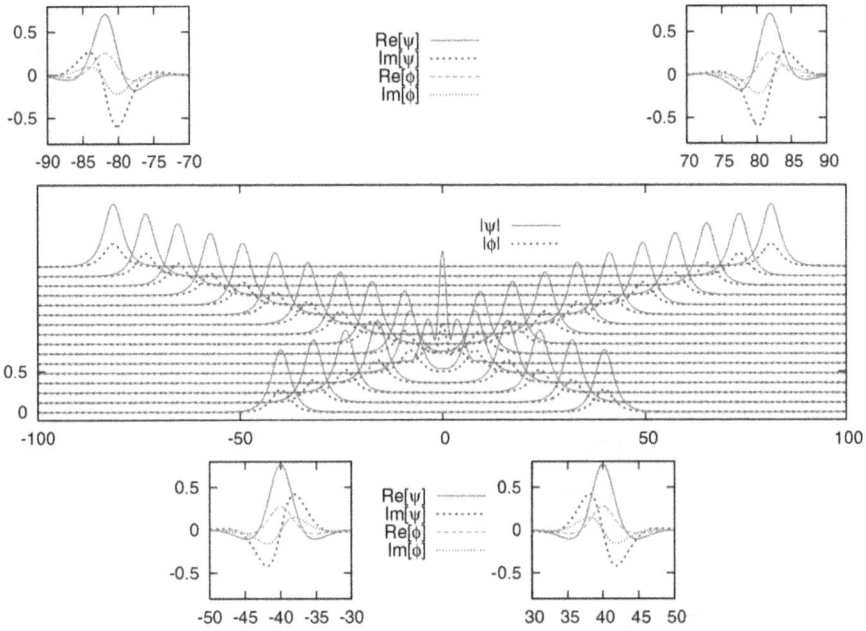

(a) $\delta_l = \delta_r = 0°$

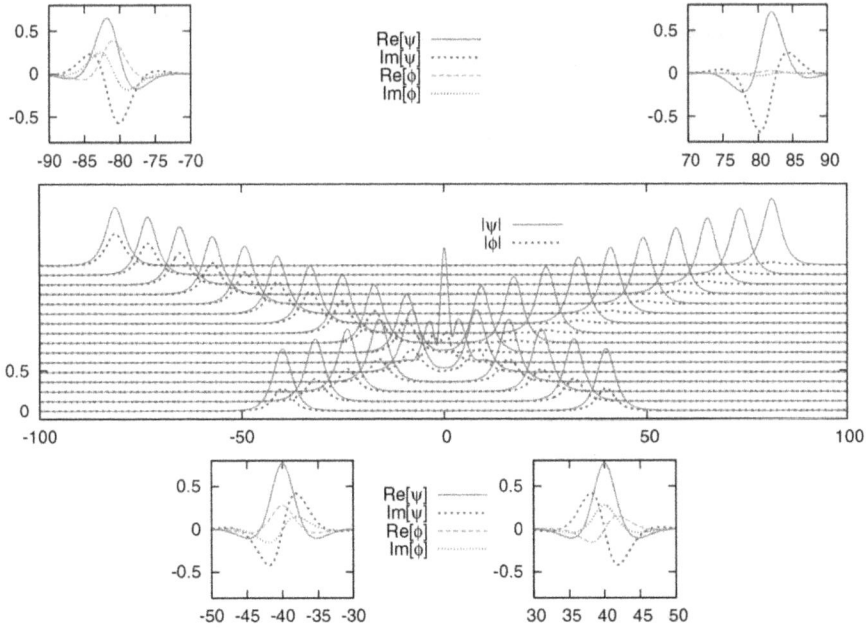

(b) $\delta_l = 0°$, $\delta_r = 90°$

Figure 2.12. Interaction of two QPs with initially equal polarizations $\theta_l = \theta_r = 20°$.

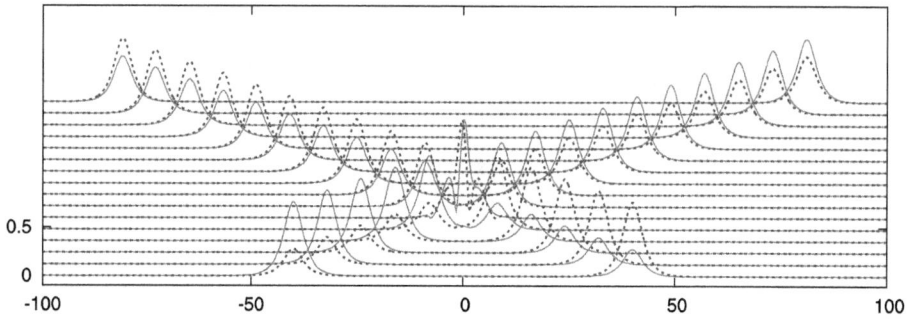

(a) $\delta_l = \delta_r = 0°$

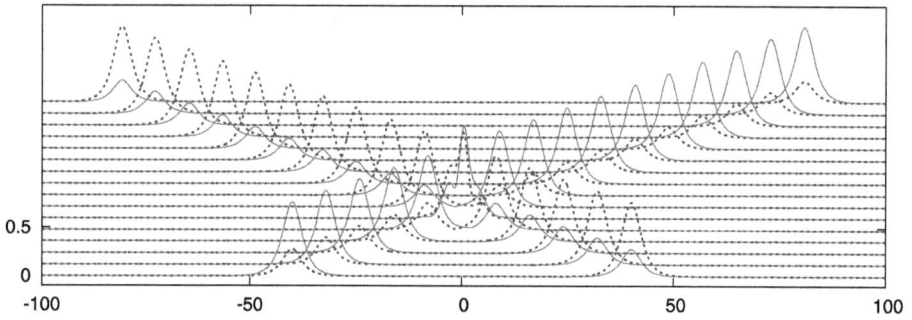

(b) $\delta_l = 0°$, $\delta_r = 90°$

Figure 2.13. Non-Manakov initial condition: two solitons of different polarization in the initial moment: $\theta_1 = 20°$, $\theta_r = 70°$.

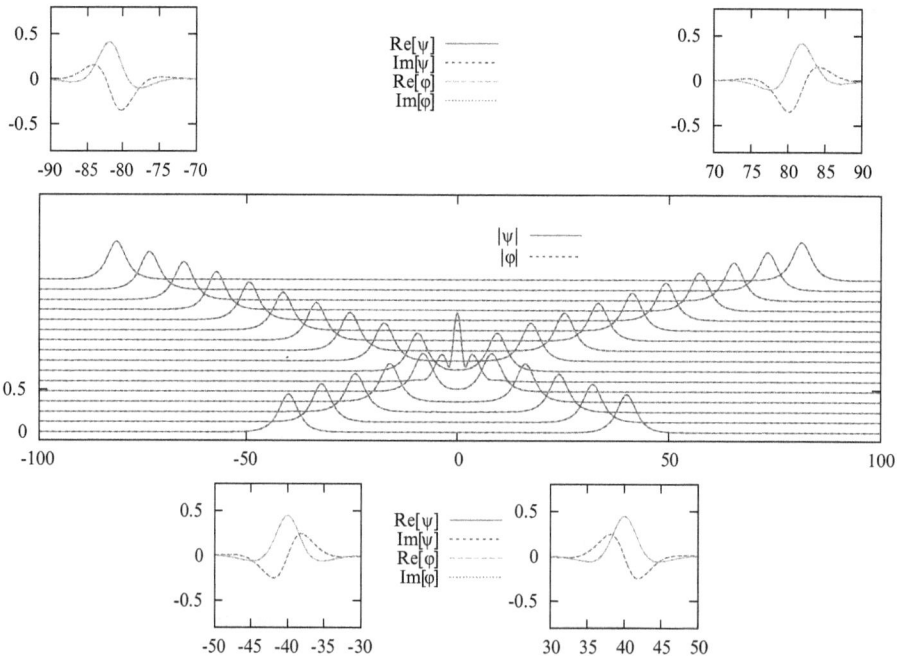

Figure 2.14. Interaction of two QPs of equal polarizations of 45° for $\alpha_2 = 0.5$ and no phase difference $\delta_1 = \delta_r = 0°$.

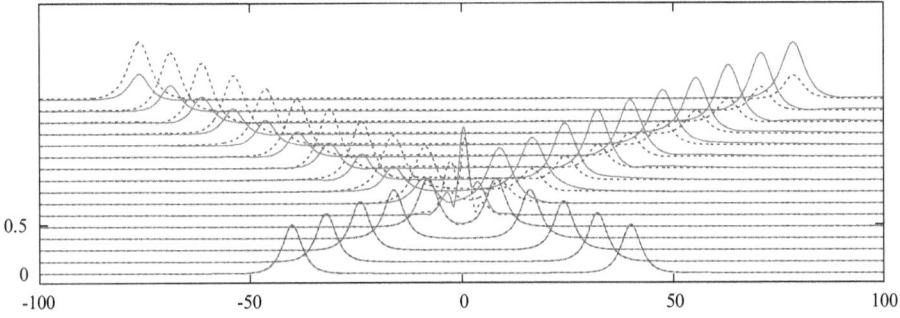

Figure 2.15. Interaction of two QPs of equal polarizations of 45° for $\alpha_2 = 0.25$ and initial phases $\delta_1 = 0°$, $\delta_1 = 90°$.

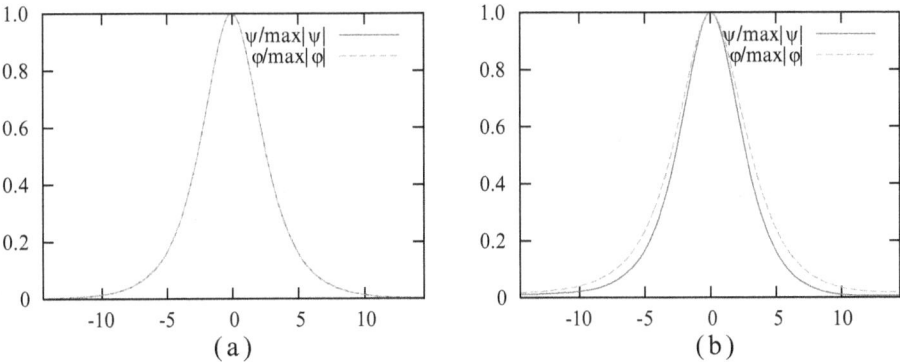

Figure 2.16. Test for the polarization of the outgoing QPs.

Conclusions

We investigate numerically the collisions of circularly polarized solitons in CNLSEs. We compose the initial conditions as a superposition of two one-soliton solutions and investigate their collision as QPs.

For the case of trivial value of the cross-modulation parameter $\alpha_2 = 0$, we investigate different combination of initial QPs. If the two initial QPs have the same polarization angle, we are able to recover the analytical two-soliton Manakov's solution only when the two real and imaginary parts of the QPs are in phase. When there is a difference between the phases of the initial QPs, the two QPs that re-emerge from the collision, have polarizations different from the initial one.

We also investigate cases when the initial polarizations of the two QPs are different. Such a case does not fall in the category of Manakov's two-soliton solution. The polarization of each QP changes after the collision.

For the case nontrivial cross-modulation parameter, we show that a solution of Manakov type can still be obtained analytically for one specific polarization: $\theta = 45°$. After the collision, two QPs with this polarization which have the same phases remain QPs with the same polarization. Contrary to the case $\alpha_2 = 0$, a zero phase leads to larger disparity of the outgoing polarizations, while a phase of 90°

preserves the polarizations. As should have been expected for a problem with nontrivial cross-modulation, the resulting QPs are not longer circularly polarized, but have a general elliptic polarization.

In all of the considered cases, we find a conservation of the total polarization.

We aim to pay a special attention of the nonlinear interaction. To this end, we study the head-on collisions of linearly polarized QPs [100]. We establish that the initial polarization on the place of interaction discontinues and becomes elliptic one. Yet, for large values of the coefficient of nonlinear coupling ($\alpha_2 > 4$), the collision bear new soliton waves (QPs) [99]. The conservative finite-difference implementation ensures the conservation of the full energy before and after the interaction. In this way, the individual energy of the main QPs after the interaction becomes negative since the high energetic excitations and radiation take away part of it. These effects shed light on the intimate mechanisms of interaction of QPs.

2.4 Impact of the large cross-modulation parameter on the collision dynamics of quasi-particles governed by vector nonlinear Schrödinger equation

Apart from its crucial importance in optics, the NLSE is a nonlinear dispersive equation in its own right, in which the nonlinearity and dispersion can be balanced allowing localized structures to propagate without change. This balance we termed the 'Boussinesq paradigm.' Elaborate models, such as CNLSE (see the extensive survey [88] and the literature cited therein), involve more parameters and have richer phenomenology but, as a rule, are not fully integrable and require numerical approaches. As we commented above, the non-fully-integrable models possess three conservation laws: for (wave) 'mass', (wave) momentum, and energy and these have to be faithfully represented by the numerical scheme. An implicit scheme of Crank–Nicolson type was first proposed for the single NLSE in the extensive numerical treatise [97]. The concept of the internal iterations was applied to CNLSE in [23] and extended in [92] in order to ensure the implementation of the conservation laws on difference level within the round-off error of the calculations. The CNLSE is also investigated numerically in [58]. Here, we generally follow the works [23, 92] but focus on a new complex-variable implementation of the conservative scheme. Our aim is to better understand the particle-like behavior of the localized waves. Let us remember that we call a localized wave a QP if it survives the collision with other QPs (or some other kind of interactions) without losing its identity.

Coupled nonlinear Schrödinger equations (CNLSEs)

In optics, the most popular model is the cubic Schrödinger equation, which describes the single-mode wave propagation in a fiber [8]. A general form of the CNLSEs reads as (2.2). We will refer to α_2 in (2.2) as a cross-modulation parameter (see also [53]) because when $\alpha_2 \neq 0$, one actually gets interaction between the two components. In fact, α_2 plays an important role in defining the elliptic, circular and linear polarizations. In the case $\alpha_2 \neq 0$, the integrability is lost, and numerical methods are to be used. The presence of the parameters, γ and Γ, makes the phenomenology of the system much richer. In particular, they allow us to study phenomena such as 'self-dispersion', 'cross-dispersion', and dissipation, etc (see [92] and the literature cited therein).

For $\Gamma = \gamma = 0$, equations (2.2) are alternatively called the Gross–Pitaevskii equations or a system of Manakov type. As already mentioned, they can be solved analytically for $\alpha_2 = 0$. We derive the pertinent theoretical and numerical formulas for the general case, but focus our attention on the case $\Gamma = \gamma = 0$ in the actual numerical experiments.

Conservation laws and Hamiltonian structure of CNLSE

It is very important for the quasi-particle approach to have the Hamiltonian structure of the system under consideration. The Hamiltonian for a single NLSE can be found in [12] (see also the monographs [1, 10]). For CNLSE, a limited version of the Lagrangian is given in [110], but no fully-fledged Lagrangian is available in the literature. For this reason, we begin with the following:

Theorem 1. The Lagrangian of the CNLSE system equations (2.2) is given by

$$
L \overset{\text{def}}{=} \frac{i}{2}(\psi_t \bar{\psi} - \bar{\psi}_t \psi) + \frac{i}{2}(\phi_t \bar{\phi} - \bar{\phi}_t \phi) - \beta(|\psi_x|^2 + |\phi_x|)
$$
$$
+ \frac{\alpha_1}{2}(|\psi|^4 + |\phi|^4) + (\alpha_1 + 2\alpha_2)|\psi|^2|\phi|^2 + \gamma(|\psi|^2 + |\phi|^2) + 2\Gamma[\Re(\bar{\psi}\bar{\phi})].
\tag{2.25}
$$

Proof Direct inspection shows that the Euler–Lagrange equations (E–L) for the conjugated functions $\bar{\psi}$ and $\bar{\phi}$ give the original equations (2.2), while E–L for the functions ψ and ϕ give the conjugated system

$$
-i\bar{\psi}_t = \beta\bar{\psi}_{xx} + [\alpha_1|\psi|^2 + (\alpha_1 + 2\alpha_2)|\phi|^2]\bar{\psi} + \gamma\bar{\psi} + \Gamma\bar{\phi}
\tag{2.26}
$$

$$
-i\bar{\phi}_t = \beta\bar{\phi}_{xx} + [\alpha_1|\phi|^2 + (\alpha_1 + 2\alpha_2)|\psi|^2]\bar{\phi} + \gamma\bar{\phi} + \Gamma\bar{\psi}.
\tag{2.27}
$$

□

Now the Hamiltonian is defined via the Legendre transformation

$$
H = \frac{\partial L}{\partial \psi_t}\psi_t + \frac{\partial L}{\partial \bar{\psi}_t}\bar{\psi}_t + \frac{\partial L}{\partial \phi_t}\phi_t + \frac{\partial L}{\partial \bar{\phi}_t}\bar{\phi}_t - L
$$
$$
= \beta(|\psi_x|^2 + |\phi_x|^2) - \frac{\alpha_1}{2}(|\psi|^4 + |\phi|^4) - (\alpha_1 + 2\alpha_2)|\psi|^2|\phi|^2 - \gamma(|\psi|^2 + |\phi|^2)
$$
$$
- 2\Gamma[\Re(\bar{\psi}\bar{\phi})].
\tag{2.28}
$$

On the interval $x \in [L_1, L_2]$, we impose trivial b.c. for the functions ψ, ϕ and evaluate the integral of the following

$\overline{\psi}_t \cdot$ equation (2.2a) + $\psi_t \cdot$ equation (2.26) + $\overline{\phi}_t \cdot$ equation (2.2b) + $\phi_t \cdot$ equation (2.27).

After some tedious but straightforward derivations in which the asymptotic boundary conditions are acknowledged, we obtain the conservation law for the energy

$$
\frac{d}{dt} \int_{-L_1}^{L_2} \left[-\beta(|\psi_x|^2 + |\phi_x|^2) + \frac{\alpha_1}{2}(|\psi|^4 + |\phi|^4) + (\alpha_1 + 2\alpha_2)|\psi|^2|\phi|^2 \right.
$$
$$
\left. + \gamma(|\psi|^2 + |\phi|^2) + 2\Gamma[\Re(\overline{\psi}\overline{\phi})] \right] dx = 0. \tag{2.29}
$$

Since the integrand in equation (2.28) is $-H$, it proves that the Hamiltonian dose represents the energy density for this system. For the wave momentum, we have the definition

$$
P = \int_{-L_1}^{L_2} \frac{i}{2} \left[(\psi\overline{\psi}_x - \psi_x\overline{\psi}) + (\phi\overline{\phi}_x - \phi_x\overline{\phi}) \right] dV = -\int_{-L_1}^{L_2} \Im\left[\psi\overline{\psi}_x + \phi\overline{\phi}_x\right] dx. \tag{2.30}
$$

Similarly to the derivations for the energy, we consider

$$
\overline{\psi}_x \cdot \text{equation (2.2}a\text{)} + \psi_x \cdot \text{equation (2.26)} + \overline{\phi}_x \cdot \text{equation (2.2}b\text{)} + \phi_x \cdot
$$
$$
\text{equation (2.27)}. \tag{2.31}
$$

and after some algebra we get

$$
\frac{dP}{dt} = \int_{-L_1}^{L_2} \frac{d}{dx} \left\{ \beta(|\psi_x|^2 + |\phi_x|^2) - \frac{\alpha_1}{2}(|\psi|^4 + |\phi|^4) - (\alpha_1 + 2\alpha_2)|\psi|^2|\phi|^2 \right.
$$
$$
\left. - \gamma(|\psi|^2 + |\phi|^2) - 2\Gamma[\Re(\overline{\psi}\overline{\phi})] \right\} dx = H|_{-L_1}^{L_2}, \tag{2.32}
$$

which is the balance law for the wave momentum. For asymptotic b.c., this balance law becomes a conservation law. Note that in higher dimensions, the wave momentum is a vector and in establishing equation (2.32), the divergence theorem has to be used. Note that in the case of asymptotic b.c., i.e. when $|L_1|$, $|L_2| \to \infty$, the balance law for the momentum, equation (2.32), becomes a conservation law as well.

In the end, we take the integral of the combination

$$
\overline{\psi}_t \cdot \text{equation (2.2}a\text{)} - \psi_t \cdot \text{equation (2.26)} + \overline{\phi}_t \cdot \text{equation (2.2}b\text{)} - \phi_t \cdot \text{equation (2.27)}, \tag{2.33}
$$

which gives us the conservation of 'mass'

$$
\frac{dM}{dt} = 0, \quad M \equiv \int_{-L_1}^{L_2} (|\psi|^2 + |\phi|^2) dx. \tag{2.34}
$$

Conservative difference scheme in complex arithmetic

Our purpose is to create a difference scheme that is not only convergent (consistent and stable), but also reflects the conservation laws. Actually, we already considered this topic in detail and we refer the reader to equations (2.8).

We mention here that for physically reasonable time increments τ, the number of internal iterations needed for convergence is four to six, which is a small price to pay to have fully implicit, nonlinear and conservative scheme.

Interaction of solitons: numerical results

CNLSE possess solutions that are localized envelopes. Since CNLSE is a two-component system, generally speaking the envelopes for each ψ, ϕ can have different amplitudes. In addition, the carrier frequency for each component can be different. This brings into view the notion of polarization which reflects the relative importance of the two components. The most general polarization is the elliptic one (see e.g. [95]) but no analytical expression is available for the solution. As an initial condition, we consider the case of a particular linear polarization when one of the components is not present (see equation (2.12)).

For consistency, we keep the grid parameters fixed and the values of the initial phase speeds are selected to be $c_{\text{left}} = 1$, $c_{\text{right}} = -0.5$. We began with the case $\alpha_2 = 0$ and confirmed that there was no interaction between the two initially orthogonal pulses. The interaction becomes appreciable around $\alpha_2 = 2$ when the smaller QP becomes elliptically polarized (see the upper panel of figure 2.17). The transformation of pulses after collision was reported in figure 3 in [66], but no systematic numerical experiments are available in the literature about the change of polarization as a result of the integration. In addition, because of the interaction, the

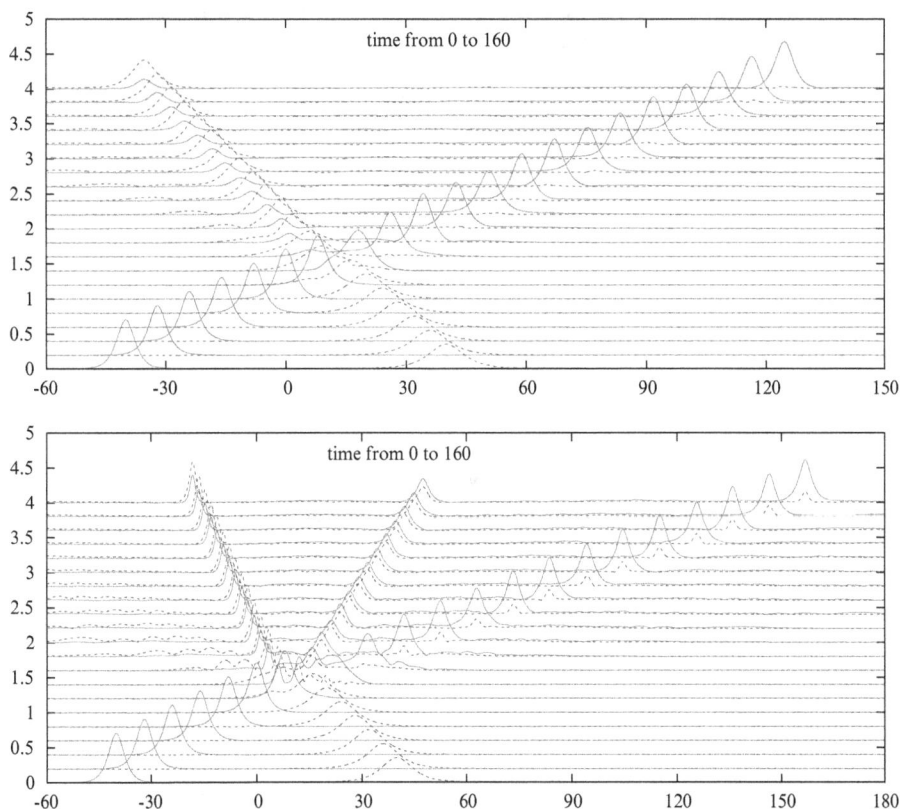

Figure 2.17. Head-on collision for $c_{\text{left}} = 1$, $c_{\text{right}} = -0.5$. Solid line: $|\psi|$; dashed line: $|\phi|$. Upper panel: $\alpha_2 = 2$, lower panel: $\alpha_2 = 6$.

larger QP is slightly accelerated, while the smaller QP is slowed down. Increasing the cross modulation to $\alpha_2 = 4$ does not change qualitatively the above described dynamics of QP. The difference from the case $\alpha_2 = 2$ is mostly in the increased amplitude of the excited orthogonal solitons, especially the ψ-soliton accompanying the smaller left-going ϕ soliton. The support of the left-going QP becomes shorter, and the QP slows further down while the larger (right-going) QP is even more accelerated.

The qualitative change of the dynamics is observed around $\alpha_2 = 6$. As shown in the lower panel of figure 2.17, apart from the increased amplitudes of the excited accompanying components, a third QP appears as a result of the interaction. There was some small hint at this effect even for smaller α_4, but now it is more pronounced. It is interesting that the third QP is moving to the right and its birth does not seem to slow the right going QP. In fact, the latter acquires an even faster phase speed. The trajectories of the centers of QPs are presented in figure 2.18. The kinetic energy of the larger QP increases after the interaction, while the internal energy of the smaller particle decreases so much that it becomes negative (to a smaller extent this effect is observed also for the weak interaction, $\alpha_2 = 2$). In a sense, the internal energy of the smaller QP is converted to the kinetic energy of the larger particle and is also used to create a new QP between the two main QPs. We can call this the 'recoil effect'.

To understand better the mechanism interaction, we investigate the motion of a QPs in their moving frames. To this end, we track the center of the QP assuming that

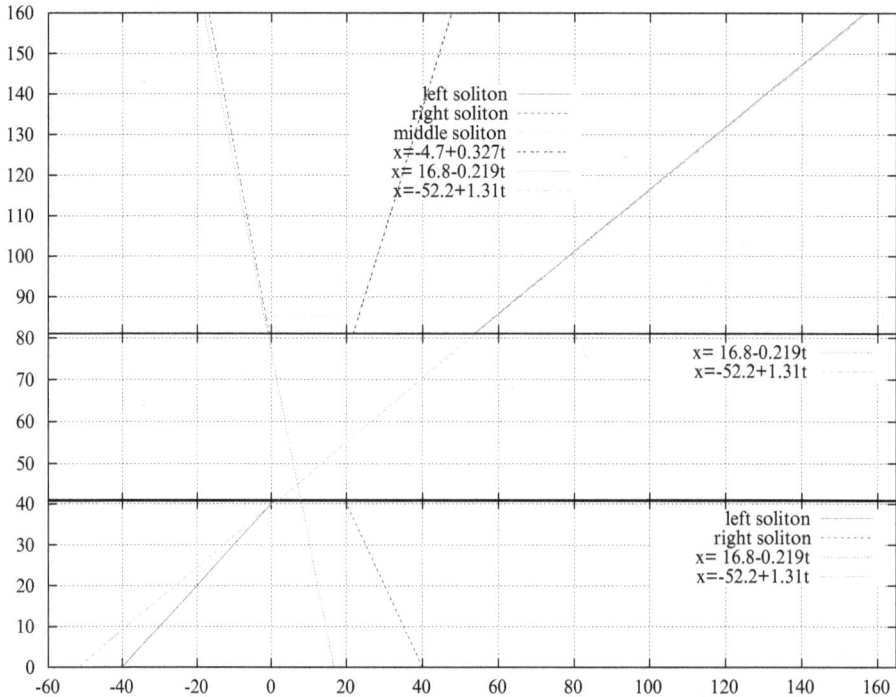

Figure 2.18. Trajectories of the quasi-particles for $\alpha_2 = 6$. Horizontal axis is x.

it is the point of maximum of the modulo of the larger component. Then, we find the time dependence of ψ and ϕ in the point of the said maximum. As a result, we get periodic functions of time whose periods give the carrier frequencies in the moving frame. The change of the carrier frequency in the wake of the interaction is depicted in figure 2.19 where only the left-going QP is considered. The situation with the right-going particle is qualitatively similar, but there are quantitative differences. For instance, the carrier frequency of the excited ϕ-component for the right-going QP is much higher. The relevant graph is not presented here in order to not overload the text.

It is interesting to note here that for the newborn QP (see the lower panel of figure 2.17), the amplitudes of the ψ- and ϕ-components are commensurate but the carrier frequency of the latter is much higher, which means that it is not circularly polarized with angle close to 45° but is, in fact, an elliptically polarized solution. The third QP has somewhat larger support than the left-going one.

We find the best fit of formulas of type of equation (2.12) to the profiles after the interaction in order to extract the pure QPs from the field after the interaction. Although in the case of elliptic polarization, the solution is not strictly a *sech*, it is still close enough in order to make this kind of extraction worthwhile. Figure 2.20 shows the results. Figure 2.20(a) presents the ψ- component of the large (right-going) QP. Since the ϕ component is not very large, we do not present it in the figure. Clearly, the deformation of the larger QP is not very strong. Figures 2.20(b) and (c)

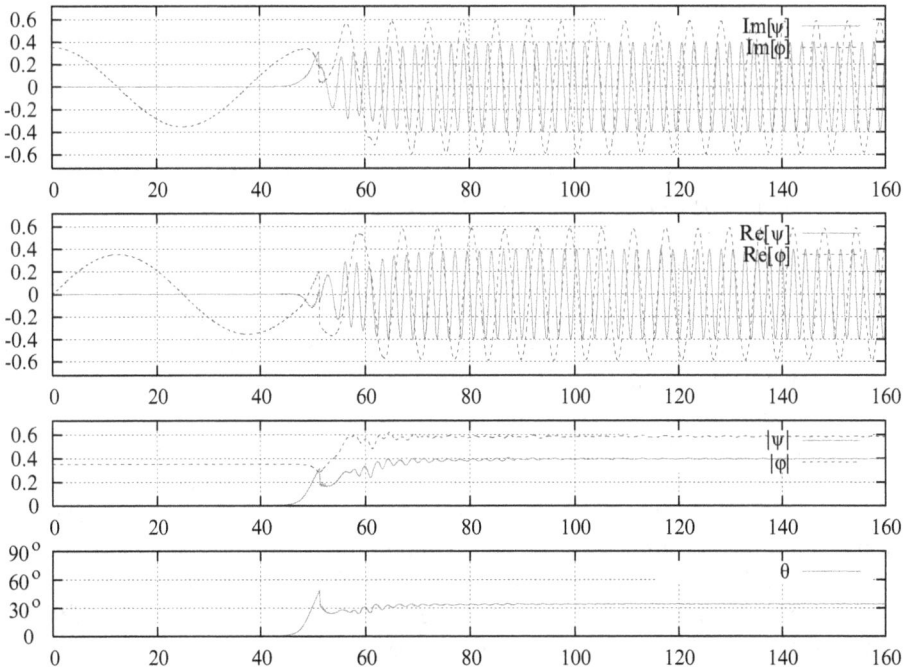

Figure 2.19. Temporal behavior of the left-going QP in its moving frame for $\alpha_2 = 6$. Horizontal axis is the time, t.

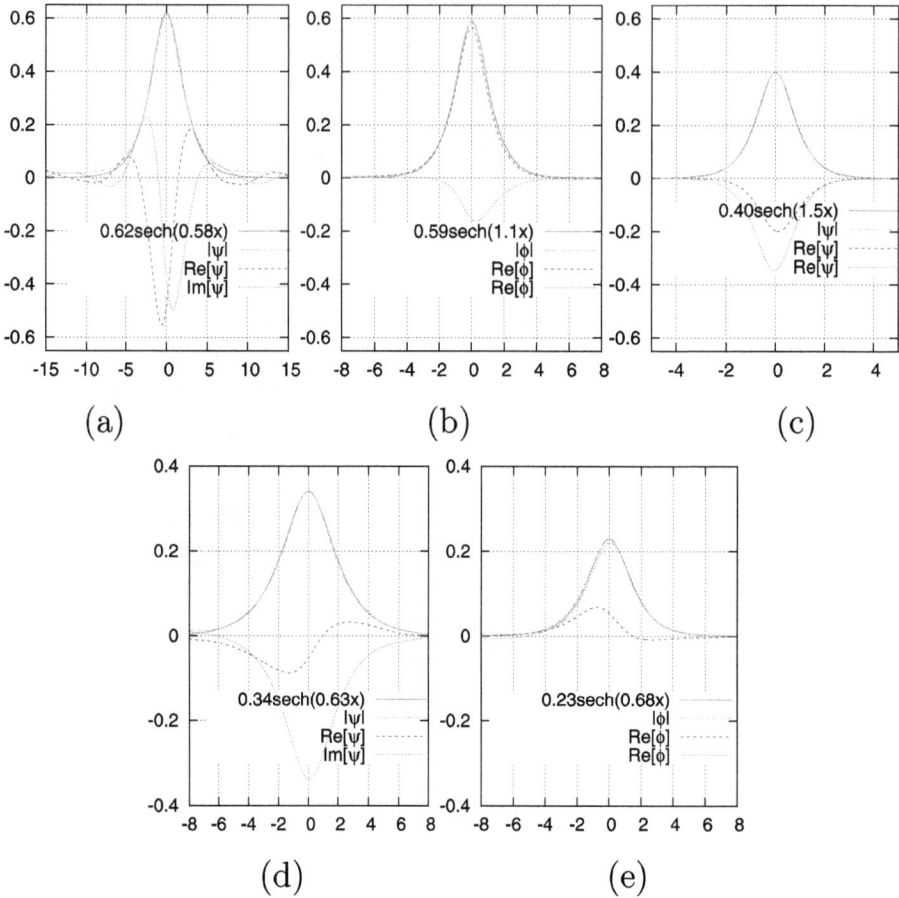

Figure 2.20. Best-fit approximations of *sech* type to the modulo of ψ and ϕ for $c_l = 1$, $c_r = -0.5$, $\alpha_2 = 6$. (a): ψ component of the right going QP; (b), (c): ϕ- and ψ- components of the left-going QP; (d), (e) ψ- and ϕ- component of the newborn QP. The horizontal axis is $x - X(t)$, where $X(t)$ is the center of the respective QP.

present the components of the smaller (left-going) QP after the collision. The excited signal is commensurate with the main component. A similar situation is observed for the newborn QP in (d) and (e). The estimated best-fit parameters can be used to analytically compute the energies and momenta of the different QPs.

One can see in table 2.2 that the energy of the left-going QP is negative, $E = -0.4020$ as computed on the basis of the best-fit parameters. This raises the question of the reliability of the result. In order to verify the latter, we clipped the region around the left-going QP and computed numerically the energy and the other characteristics directly from the available profile. We have found that the directly evaluated energy over the truncated interval is $E = -0.3878$, which confirms the validity of the best-fit formulas. Respectively, for the directly computed mass, we got $M = 0.413$, which is in very good agreement with the best-fit result of 0.4231 (see the respective entry of table 2.2).

A natural question arises here about the symmetry of the interaction, when the initial configuration of the solitons is perfectly symmetric. To investigate this, we conducted a numerical experiment with two QPs with equal initial phase speeds, $c_l = -c_r = 1$. The expectation is that the third QP will stay in the origin of the coordinate system, i.e. a standing soliton should be born. Indeed, figure 2.21 shows that our computations confirmed this supposition. The two main QPs evolve perfectly symmetrically after the collision and the newborn QP is a standing soliton. Note that the amplitudes of the two components of the standing soliton appear to be equal to each other. A close examination of their carrier frequencies shows that in fact they are also equal, which means that the standing soliton has circular polarization with angle $\theta = 45°$. The masses, energies and momenta of the QPs in the symmetric case are qualitatively similar to the case $c_l = 1$, $c_r = -0.5$ and will not be presented here.

In the end, we consider also the case $\alpha_2 = 10$ when the cross-modulation becomes a very strong effect. Figure 2.22 shows the further increase of the excitability of the

Table 2.2. Properties of quasi-particles (QPs) for $c_l = 1$, $c_r = -0.5$, and $\alpha_2 = 6$ before and after the collision.

Soliton	M	c	E_k	E_p	E	P
			Before collision			
Left (right going)	1.0	1.0	0.5000	−0.1667	0.3333	1.00
Right (left going)	0.5	−0.5	0.0625	−0.0208	0.0417	−0.25
Total	1.5	0.5[a]	0.5625	−0.1875	0.3750	0.75
			After collision			
Right (right going)	0.6830	1.31	0.5861	−0.1848	0.4013	0.8948
Left (left going)	0.4231	−0.219	0.0102	−0.4122	−0.4020	−0.0927
Middle (newborn)	0.2450	0.327	0.0131	−0.0797	−0.0666	0.0801
Total	1.3512	0.6529[a]	0.6093	−0.6767	−0.0673	0.8822

[a]The speed of the center of mass $c = P_{total}/M_{total}$

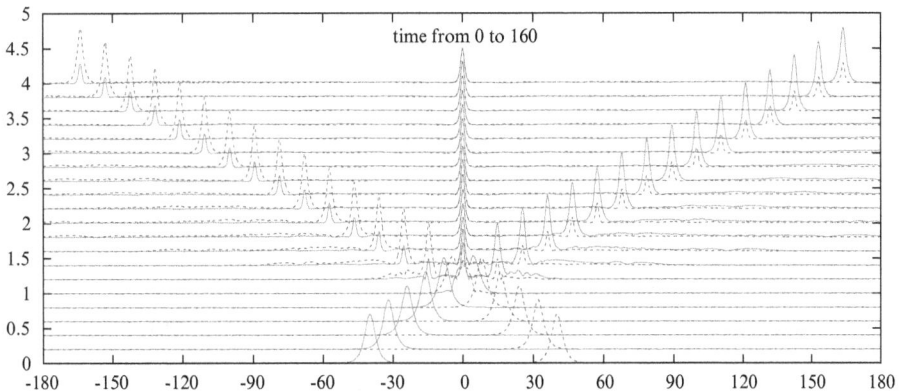

Figure 2.21. Collision of two equal QPs for $\alpha_2 = 6$, $c_l = -c_r = 1$.

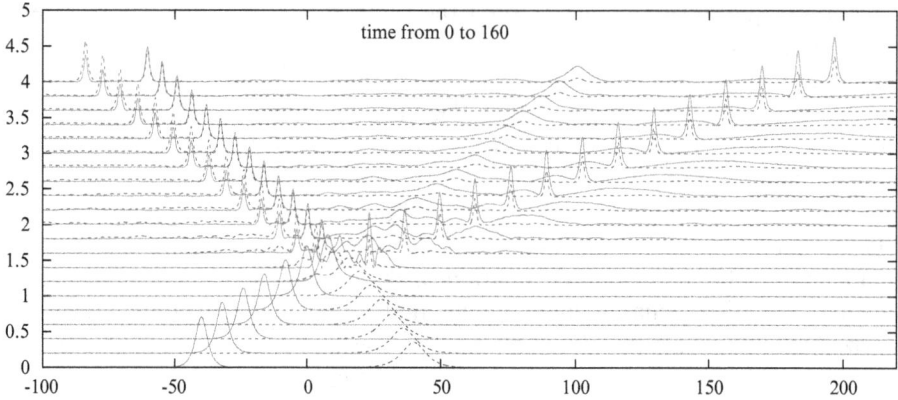

Figure 2.22. Collision of two equal QPs for $\alpha_2 = 10$, $c_1 = 1$, $c_r = -0.5$.

Table 2.3. Integral characteristics of excitations.

	Left forerunner	Between QP-2 and QP-3	Between QP-3 and QP-4	Right forerunner
M	0.0431	0.0269	0.05908	0.1969
P	0.0082	0.0083	0.07669	0.4087
E	0.2215	0.0044	0.0045	0.7225

system. The main difference from $\alpha_2 = 6$ is that the left-going QP splits into two very narrow QPs, which are moving with much faster phase speed than the original left-going QP.

In addition, the right-going QP also gets its support shortened. Another very important trait of the interaction for this very large $\alpha_2 = 10$ is the appearance of considerable radiation that carries away the kinetic energy of the system and the total energy of the QPs is radically different from the total energy of the initial wave profile. The differences are so drastic that the sum of QPs energies can even become negative. This means that the energy was carried away by the radiation.

In order to confirm this conjecture, we consider the final time stage from figure 2.22, clip away the main QPs and calculate the energy of the wave profile that is left in the reduced region. As one can see, the amplitude of the radiation that dwells in the mentioned region is rather small. Yet the energy of the oscillations is very large, particularly, the kinetic energy. We enumerate the different QPs from left to right, namely the leftmost soliton is QP-1, and the rightmost is QP-4. The results are organized in table 2.3 for the different regions, save the small interval between QP-1 and QP-2, whose characteristics are negligible.

The masses and the pseudomomenta of the excitations are commensurate with their relative importance for the amplitudes of the total wave profile. The predominant part of the kinetic energy is concentrated in the left and right forerunners because they propagate with very large phase speeds, and span large portions of the region.

Finally, we add together the energies of all four QPs and the excitations alike and get results, which are within 20% of the total energy as conserved of the scheme. This means that the large negative energies of the QPs are the result of the evacuation of energy by the forerunners and is not a numerical effect. This kind of transformation is a physical effect that is connected with the excitability of the system and is not present in the case of a single Schrödinger equation.

Conclusion

We develop a complex-arithmetic implementation of a conservative difference scheme for CNLSEs. To this end, a special solver for Gaussian elimination with pivoting is developed for multi-diagonal complex-valued matrices, which is a generalization of the solver created earlier by one of the authors. The new solver allows us to use grids of considerable sizes (up to 20000 grid points) and small time increments obtaining thus a reliable approximation.

The algorithm is validated by the mandatory numerical tests involving doubling the mesh size and halving the time increment, as well as by the direct comparison with the real-arithmetic schemes [23, 92]. The advantages of the new scheme are that the band of the matrix is twice as small and that the overall number of unknowns is also twice as small. Our numerical tests show that, as expected, it performs four times faster than the scheme from [23, 92].

The new tool developed here allow us to consider physically important sets of parameters of the CNLSE and to investigate the role of nonlinear coupling in the quasi-particle dynamics. The nonlinear coupling results in transforming the original circular polarizations of the two QPs to generally elliptic ones. This means that although the initial conditions in each of the initial locations are nontrivial for only one of the functions, after the interaction both functions acquire nontrivial amplitudes for all QPs that emerge from the collision.

We consider as an initial profile the superposition of solitons with linear polarizations, one of them having only a ψ-component, and the other—only a ϕ-component. Then, this initial profile is allowed to evolve according to the CNLSE system. For $\beta = 1$, $\alpha_1 = 1$, $\gamma = \Gamma = 0$, the results of the collision depend solely on the nonlinear coupling parameter α_2 (cross-modulation parameter). We have found that for $\alpha_2 < 4$, the collision of the two initial solitons with linear polarizations results in two solitons again, but with different polarizations. The smaller soliton suffers more from the interaction and its polarization becomes elliptic.

For moderate values of the cross-modulation parameter, $\alpha_2 = 6$, we have found that an additional QP is born which propagates in the direction of the faster QP, while the initially smaller QP considerably decelerates. The initially faster QP becomes even faster. The effect of the nonlinearity is so profoundly felt that even the faster QP becomes elliptically polarized, although to a smaller extent than the slower and new-born QPs. If the initial QPs have the same phase speeds, the new-born QP is a standing soliton with circular polarization of 45°. We have observed that the energies of the initially slower QP and the new-born QP become negative after the interaction. This is due to the fact that they are steeper (with smaller support) and the potential energy (which is the stored elastic energy) becomes very negative and dominates the kinetic energy. This means that part of the energy is taken away after the interaction by the excited radiation.

Finally, we consider the case of large value of cross-modulation parameter, $\alpha_2 = 10$ and find that the dynamics changes even more radically. Now, two new QPs are born, accompanying the two initial QPs which are also radically transformed in the sense of polarization and energy. All four QPs have elliptic polarizations and negative total energies. In addition, fast forerunners with large positive (mostly kinetic) energies are born which preserve the total energy of the system, i.e. the radiation in form of forerunners evacuates positive energy leaving the ensemble of QPs with negative total energy.

2.5 Repelling soliton collisions in vector nonlinear Schrödinger equation with negative cross modulation

A new kind of repelling collision is discovered for negative values of the cross-modulation coupling parameter, α_2 in equations (2.2). The results show that as the latter becomes increasingly negative, the behavior of the solitons during the interaction change drastically. While for $\alpha_2 > 0$, the solitons pass through each other, a negative threshold value $\alpha_2^* < 0$ is found below which the solitons repel each other [93, 94].

The system, equations (2.2), is alternately called the Gross–Pitaevskii equation or an equation of Manakov type. It was derived independently by Gross [49, 51] and Pitaevskii [84], to describe the behavior of Bose–Einstein condensates as well as optic pulse propagation. Let us remind you that the functions ψ and ϕ in equations (2.2) have various interpretations in the context of optic pulses. These include the amplitudes of x and y polarizations in a birefringent nonlinear planar waveguide [17, 82, 83] and the amplitudes of x and y linear polarizations in a birefringent fiber [15, 17]; pulsed wave amplitudes of left and right circular polarizations [11, 19, 111] and pulsed wave amplitudes of symmetric and antisymmetric modes in a nonlinear coupler [59, 91].

The parameter of interest will again be the cross-phase modulation α_2, which defines the integrability of equations (2.2). We are not concerned with the linear coupling and set $\gamma = 0$ and $\Gamma = 0$.

The term proportional to $\hat{\alpha} = \alpha_1 + 2\alpha_2$ can be seen as an interaction potential between evolving pulses in the CNLSE. We intend to vary this interaction potential to uncover a new collision behavior, namely repulsion. The interaction of two solitons in CNLSE is a paragon example for interaction. The adiabatic interaction of N solitons was described in [31, 109].

We focus on the collision properties of solitons that initially have linear polarization, namely

$$\begin{pmatrix} \psi_l(x, t) \\ \phi_l(x, t) \end{pmatrix} \overset{\text{def}}{=} \begin{pmatrix} \Omega(x, t; c, n, X) \\ 0 \end{pmatrix} \quad \text{and} \quad \begin{pmatrix} \psi_r(x, t) \\ \phi_r(x, t) \end{pmatrix} \overset{\text{def}}{=} \begin{pmatrix} 0 \\ \Omega(x, t; c, n, X) \end{pmatrix}$$

where the subscripts l and r denote left or right initial QP. Respectively, Ω is the one-soliton solution of the NLSE, given by equation (2.12).

We set the goal to unearth the effect of negative cross-modulations, i.e. $\alpha_2 < 0$ on the properties of the soliton collisions in the NLSE. To this end, we focus our attention on interacting solitons with flat boundary conditions, i.e. bright solitons.

Numerical scheme

The system we study is non-integrable and a numerical approach must be used to find approximate solutions of equations (2.2).

Here, we would cite earlier numerical works including the pioneering effort [29] where a highly accurate spectral scheme was developed (see also [28] for the KdV equation (shallow water)). For water of infinite depth, a numerical scheme was developed in [68]. A decisive advance was made in [2, 3] where the schemes were designed in a manner better suiting inverse scattering techniques. The scheme used in the present work reflects the system's dynamic properties: specifically, we construct a scheme which mimics the integral constants of the system. It is a descendent of [23], and differs from [96, 97], because it makes use of internal iterations, and because it is about the CNLSE (also known as VNLSE). The latter does not permit us to perform a head-to-head comparison with the above mentioned schemes, but the performance of our scheme was validated by all standard tests in numerical analysis: doubling and halving the spacing and the time increment; increasing and decreasing the cut-off of the spatial interval emulating the infinite region, etc. Conservation of energy is especially important in the context of long-time calculations, where non-conservative methods can produce spurious data for the energy. The scheme of [23, 92] has recently been extended to a fully complex version [101, 105].

Results and discussion

We call a collision after which the original shapes re-emerge nondeformed 'elastic', while the inelastic is a collision in which additional oscillations are excited in the place of the bygone collision. A predominant part of the research in the literature deals with the positive values of the cross modulation, when the solitons pass through each other during the collisions. In optical applications, this kind of behavior is important and it seems reasonable to limit the investigation to positive cross-modulations. When investigating the quasi-particle behavior of the CNLSE solitons, one must focus on their interactions. In order to understand the analogy to the real particles, one needs to find a model that predicts repelling of the solitons after a collision. In [93, 94], it is discovered that the solitons of the CNLSE with negative cross-modulation parameters can be repelled during collision. In this section, we present our results on this new phenomenon.

As already mentioned we set $\gamma = 0$, $\Gamma = 0$. We also fix for simplicity the diffusion coefficient to $\beta = 1$, and the self-focusing to $\alpha_1 = 0.25$. For the bulk of numerical experiments, we set the two phase speeds of the left and tight solitons to $c_1 = 0.15$, and $c_r = -0.15$, respectively. However, in order to have distinguishable QPs, we chose different carrier frequencies for them, namely $n_1 = 0.03$ and $n_r = 0.1$ for the above set of phase velocities.

As already mentioned above, an analytical solution [40] is available for $\alpha_2 = 0$. Without being formally decoupled for $\alpha_2 = 0$, the CNLSE system behaves similarly to a decoupled system. As shown in figure 2.23(a), our computation confirmed this fact. The other case when the system is decoupled is for $\alpha_2 = -\frac{\alpha_1}{2} = 0.125$ and we did verify that there is no interaction whatsoever between the two QPs as the system is completely decoupled (not even a phase shift). In the red color (solid line) is shown the modulo of component ψ, while $|\phi|$ is presented in blue (dashed line). Since the initial polarizations are linear, we have only ψ component for the left soliton, and only ϕ-component for the soliton on the right. In a case where one of the components should excite the other one during the interactions (see e.g. [99]), after the collisions the two solitons would exhibit both red- and blue-color lines. In the first and last moments of time under consideration, the real and imaginary parts of ψ, ϕ are presented in additional colors (or different length of dashes and dots).

Now, if α_2 becomes less negative than $-\frac{1}{2}$, and eventually even positive, we will be in the reasonably well investigated realm of crossing interactions of the solitons. For reference, we present in figure 2.23(a) the case $\alpha_2 = 0$ for two solitons whose polarizations differ by 90°. Unlike the case $\alpha_2 = -\frac{1}{2}$, there is interaction between the QPs, but it is exactly as expected for an integrable case: the only manifestation of nonlinearity is the phase shift.

The absence of a cross-modulation manifests itself through the fact that the orthogonal components are not excited during the interaction. It was shown, e.g. in [99] for $\alpha_2 > 0$, that the orthogonal components are indeed excited after the interaction of two initially linearly polarized pulses. It is interesting to observe here in figure 2.23(b) that in our parameter range, taking $\alpha_2 = -0.1$ does lead to some excitation of an orthogonal component, but only for the taller (and slower moving) QP. In a sense, the faster QP was split into a reflected and refracted part, and the reflected part of it moves together with the other QP. It will take a special investigation to understand the role of the negative cross modulation when $\alpha_2 \in [-\frac{1}{2}\alpha_1, 0]$, because there are no cross-excitations that take place for the two boundary values in this interval, and the excitation for the rest of the values from that interval, are not similar to the case of positive cross-modulation.

Our purpose is to investigate the regimes when $\alpha_2 < -\frac{1}{2}\alpha_1$ or $\hat{\alpha} = \alpha_1 + 2\alpha_2 < 0$. We discover that decreasing α_2 down from $-\frac{1}{2}\alpha_1$ leads to the increase of the amplitude of the reflected part of the faster QP and decrease of the refracted part. In addition, the excited orthogonal mode does not stick to the slower QP, but rather speeds away from it. Figure 2.23(c) shows the case $\alpha_2 = -0.144$ when the reflected and the refracted part have approximately equal amplitudes. At $\alpha_2 = -0.157$, all traces of the refracted part disappear. Actually, for the values of the parameters considered here, the taller QP virtually stops and reverses its trajectory for $\alpha_2 = -0.15$, as shown in figure 2.24(a). For this value of α_2 there is still a small refracted part, but no excitations are observed. Finally, in figure 2.24(b), we present a genuinely repelling interaction for $\alpha_2 = -0.2$. One can see that there is no excitation of the orthogonal components. The physical meaning of this observation is that a repulsive component appears in the potential of

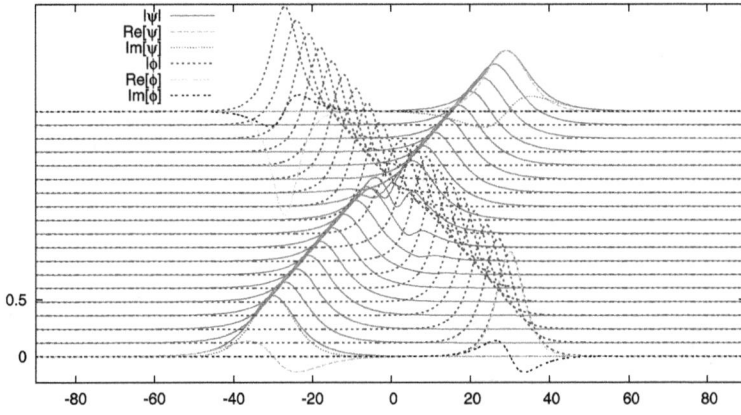

(a) $\alpha_2 = 0$, $\hat{\alpha} = 0.25$

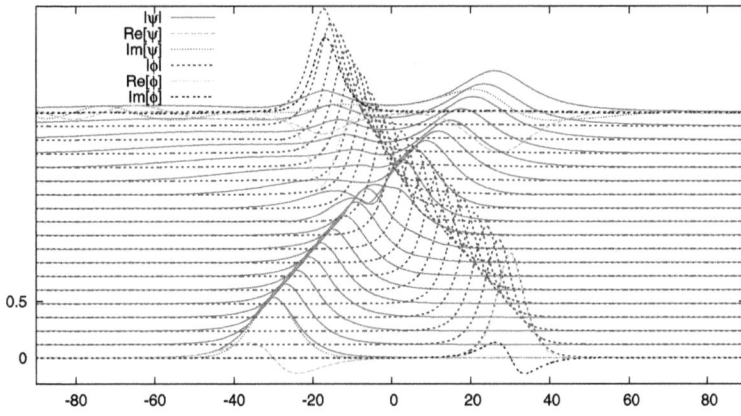

(b) $\alpha_2 = -0.1$, $\hat{\alpha} = 0.05$

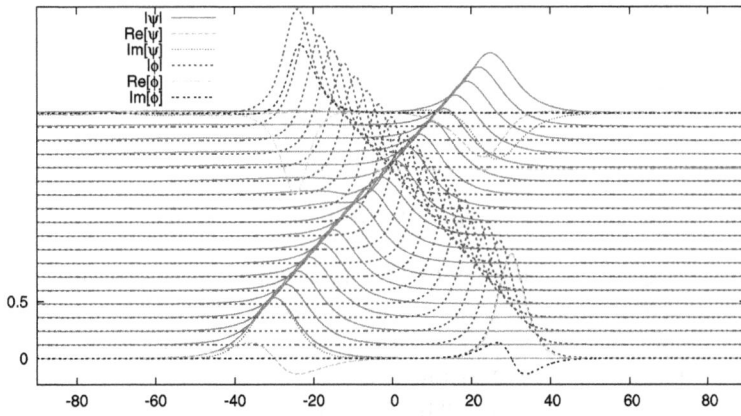

(c) $\alpha_2 = -0.12$, $\hat{\alpha} = 0.01$

Figure 2.23. Crossing collisions for $c_l = 1 = -c_r$, $n_l = 0.03$, $n_r = 0.1$.

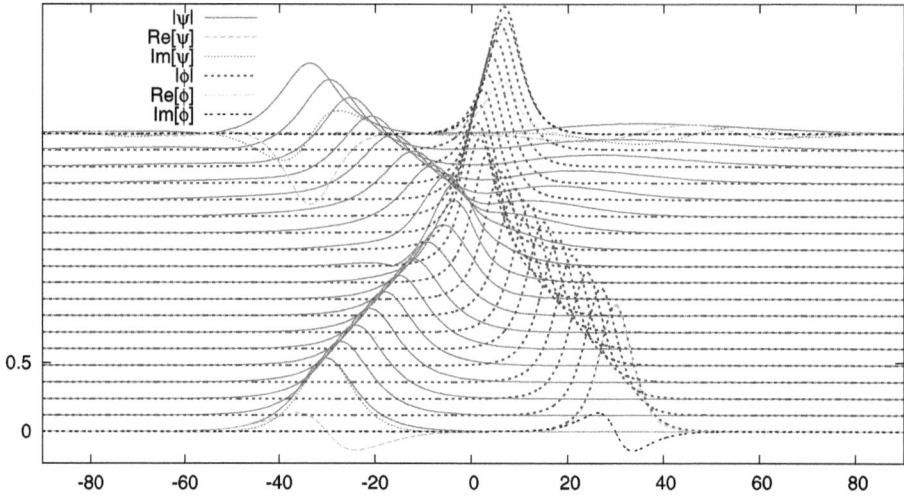

(b) $\alpha_2 = -0.15$, $\hat{\alpha} = -0.05$

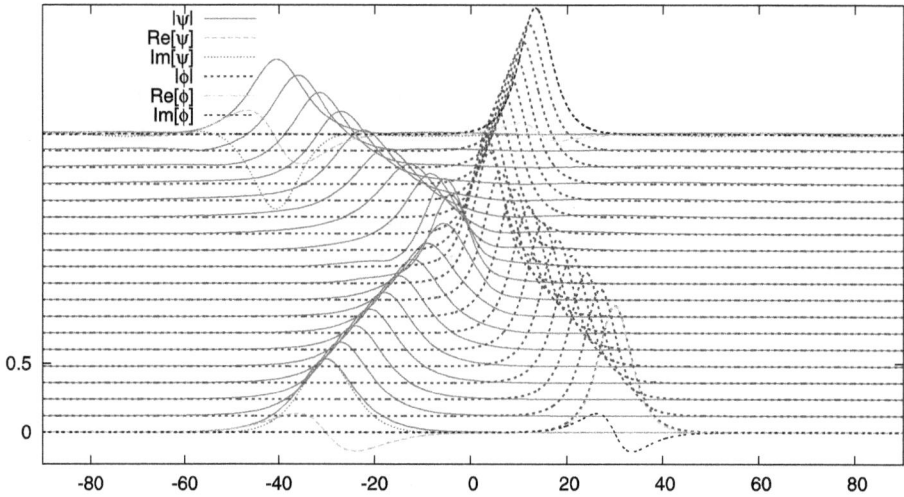

(c) $\alpha_2 = -0.194$, $\hat{\alpha} = -0.138$

Figure 2.24. Repelling interactions for negative values of the overall coupling parameter $\hat{\alpha} = \alpha_1 + 2\alpha_2$.

interaction of the QPs. The interaction potential becomes so negative that the QPs stop and turn back before they can get close enough to overlap. Thus, the non-linearity does not have the chance to excite the respective orthogonal mode. This is an important new result which is not very intuitive. Another model in which repulsion can take place is the sine-Gordon equation (sG), when the soliton–soliton interactions are investigated. The soliton–antisoliton interactions in sG are attractive. For the time being, we have not implemented the coarse-grain approximation of [22] to the quasi-particles of this work, so we do not have an approximate analytical expression for the interaction potential, but the evidence unequivocally points out towards the presence of repulsive terms in the potential.

The most important observation here is that the trajectory of the smaller soliton (the one on the left has a smaller amplitude because of the smaller carrier frequency) bends more and its final trajectory demonstrates that it is moving at higher phase speed than the one with which it entered the collision. In this range of parameters, the trajectory of the larger quasi-particle appears as it is less bent and it returns with slower phase speed than its incident speed. A most interesting result transpires from our results, namely that in the repelling collisions, there are no cross-excitations of the two components (modes), which appears to be a novel result shedding light on the soliton interactions for negative cross-modulations.

In order to confirm that the above reported behavior is authentic of the system, we also investigate a case with different initial carrier frequencies and different phase speeds. We take somewhat larger carrier frequencies, which make the left particle larger. For the specific numbers chosen, the heights of the two QPs are closer to each other. It is shown in figure 2.25 that, once again, the qualitative nature of the interaction is the same: the QPs reflect completely without trace of refracted parts, or of excited signals. Because the larger QP is now not that much taller than the smaller one, the former acquires a larger phase speed after the reflection. Respectively, the smaller QP is now slightly slower after the interaction. These are the quantitative differences, but the main qualitative feature is the same: a repelling interaction takes place for a negative α_2.

Quasi-particle characteristics

In the precedence, we discover strictly repelling QP collisions in CNLSE for negative cross-modulations. Upon collision, not only does the sense of motion change, but the magnitudes of the phase speeds assume different values too. In the crossing

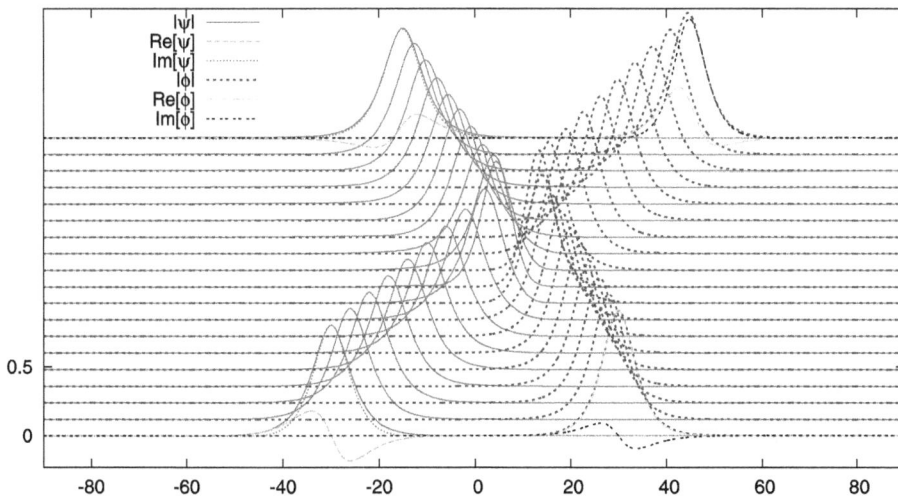

Figure 2.25. Repelling interactions for $\alpha_2 = -0.4$ and $c_l = 0.2$, $n_l = 0.07$, $c_r = -0.1$, $n_r = 0.1$.

interactions, the solitons return to their original phase speeds. Since it is not obvious how to quantify the phase shifts in this case, we will not dwell on this point here. Note here that due to the conservative nature of the numerical scheme, the total mass, energy, and pseudo-momentum of the system are conserved, despite the exotic nature of the collisions. In order to elucidate the quasi-particle aspects of the soliton collision in CNLSE, we compute the masses of the two QPs, according to equations (2.7) and (2.14). We compute the phase velocities from the trajectories of the centers of QPs after fitting lines to them in the regions far ahead and far after the collision.

Computing the masses and phase speeds of the two QPs before and after the collision enables us to identify the particle-like properties of the collisions. We present the thus computed properties in table 2.4. The upper part of the table deals with the initial values (superscript i), while the lower portion refers to the final values (superscript f). Letter M stands for mass, and c_l, c_r stand for the phase velocities of the left and right QP.

The novel result of the present work is that for negative enough cross-modulations, a repelling type of interaction takes place, making the analogy between the solitons (QPs) and the real particles much more conspicuous. Then, it is only natural to examine the pseudo-Newtonian behavior of the QPs, e.g. to compute their celerities, and pseudomomenta before and after the collision. In the case of repelling collisions, we are aided by the fact that there are no cross-excitations of the components.

Table 2.4 shows the main characteristics of the QPs before and after the collision, One sees that the individual masses and phase speeds change during the interaction, which leads to different pseudomomenta. Yet, the total pseudomomentum of the system of two QPs is conserved with a very high degree of accuracy. In this sense, one is faced with Newton-like law for conservation of the momentum of a system of two quasi-particles.

Table 2.4. Quasi-particle characteristics.

				Before collision			
α_2	M_l	M_r	c_l^i	c_r^i	$P_l^i = M_l c_l^i$	$P_r^i = M_r c_r^i$	$P = P_l^i + P_r^i$
-0.2	0.2498	0.4915	0.1498	-0.1495	0.0374	-0.0735	-0.0361
-0.25	0.2498	0.4915	0.1498	-0.1495	0.0374	-0.0735	-0.0361
-0.3	0.2498	0.4915	0.1498	-0.1495	0.0374	-0.0735	-0.0361
-0.4	0.2498	0.4915	0.1498	-0.1495	0.0374	-0.0735	-0.0361
$-0.4^\dagger$	3.919	4.996	0.1994	-0.0997	0.7814	-0.4981	0.2833
				After collision			
α_2	M_l	M_r	c_l^f	c_r^f	$P_l^f = M_l c_l^f$	$P_r^f = M_r c_r^f$	$P = P_l^f + P_r^f$
-0.2	0.2498	0.4915	-0.2425	0.0495	-0.0606	0.0243	-0.0362
-0.25	0.2498	0.4915	-0.2427	0.0503	-0.0606	0.0247	-0.0359
-0.3	0.2498	0.4915	-0.2429	0.0504	-0.0607	0.0248	-0.0359
-0.4	0.2498	0.4915	-0.2433	0.0506	-0.0608	0.0249	-0.0359
$-0.4^\dagger$	3.919	4.996	-0.1350	0.1628	-0.5291	0.8133	0.2843

$\dagger$ Case from figure 2.25

Conclusions

We consider a system of two Schrödinger equations coupled only through the nonlinear terms. We make use of a previously developed implicit conservative difference scheme. We fix all other parameters of the system, and focus on the case when the coupling parameter α_2 (called the cross-modulation parameter) can become negative, and more specifically, when $\alpha_1 + 2\alpha_2 < 0$, where α_1 is the self-focusing parameter. We obtain results for a range of values of α_2. The system is completely decoupled for $\alpha_2 = -\frac{1}{2}\alpha_1$, and behaves as a single equation for $\alpha_2 = 0$, and we recover the expected behavior numerically. Decreasing α_2 from $-\frac{1}{2}\alpha_1$ to more negative values brings into view a completely new kind of behavior of the solitons upon collision: they repel each other instead of passing through one another as is the case for positive cross-modulations.

The soliton waveforms are seen to survive the collision undeformed, despite the fact that the system cannot be proved to be fully integrable. For α_2 sufficiently negative, the collisions become fully repelling, while the total energy is strictly conserved. We compute the masses and the pseudomomenta of the QPs before and after the collision and show that the total mass and the total pseudomomentum of the system of the QPs are conserved upon collision. The repelling interactions found in this work resemble one-dimensional ballistic collisions.

Obviously, the initial polarization takes an important role in the soliton dynamics. The investigations up to now are subject to the linear and circular initial polarization of the soliton envelopes. This conjecture provides an analytic initial condition of *sech* (hyperbolic secant) kind. On the other side, we are interested in dynamics of solitary waves with arbitrary (elliptic) initial polarizations. Unfortunately, such a kind of analytic functions are not known. To surmount this advantage, we ask a solution of VNLSE that is not necessarily a sech-shaped function [102]. Under this assumption, we derive a conjugate bifurcation system of two nonlinear ordinary differential equations. We treat this system with a flat boundary and vary one of the parameters (for instance, the carrier frequency). In this way, we can generate approximate initial conditions with arbitrary polarization, the angle θ ($0° < \theta < 90°$). The resulting family of solutions contain the cases of circular and linear polarization as well.

2.6 On the solution of the system of ordinary differential equations governing the polarized stationary solutions of vector nonlinear Schrödinger equation

Consider again the CNLSE:

$$i\psi_t = \beta\Delta\psi + [\alpha_1|\psi|^2 + (\alpha_1 + 2\alpha_2)|\phi|^2]\psi$$
$$i\phi_t = \beta\Delta\phi + [\alpha_1|\phi|^2 + (\alpha_1 + 2\alpha_2)\alpha_2|\psi|^2]\phi$$

where β is the dispersion coefficient, α_1 describes the self-focusing of a signal for pulses in birefringent media, and α_2 governs the nonlinear coupling between the equations. For further convenience, we consider the case $\beta = 1$. All other cases can be reduced to this one by means of simple scaling. CNLSE possess solutions that are localized envelopes (see, for example, [72, 114]). It is easy to see that if one of the components, say ϕ, is trivially equal to zero, then the system reduces to a single scalar NLSE for the other function, ψ. Such a solution of CNLSE is called 'linearly polarized'. The *sech*-solution of the single NLSE, say for ψ-profile, is given by

$$\psi(x,\,t) = A_\psi \, \text{sech}[b_\psi(x - X - c_\psi t)] \exp\left\{ i \left[\frac{c_\psi}{2\beta}(x - X - c_\psi t) - n_\psi \right] \right\}$$

$$b_\psi^2 = \frac{1}{\beta}\left(n_\psi + \frac{1}{4\beta}c_\psi^2 \right); \quad A_\psi = b_\psi \sqrt{\frac{2\beta}{\alpha_1}}; \quad u_c = \frac{2n_\psi \beta}{c_\psi},$$

(2.35)

which means that for given phase speeds, c_ψ, and carrier frequencies, n_ψ, the solution of the above type is fully specified. The localized solution is identified by the presence of a spatial point, $x = X - c_\psi t$ ('center') which moves with a given phase speed c_ψ. The same initial relationships hold for the ϕ-profile when $\psi = 0$. The center is common for both components ψ and ϕ. The above type of analytical initial condition evidently requires the restriction $4\beta \, b_{\psi,\,\phi} > -c_{\psi,\,\phi}^2$. It should be noted here that, for the Manakov case [72] when $\alpha_2 = 0$, a solution of type of equation (2.35) can be found for circular polarization of kind $\psi(x,\,t) = \chi(x,\,t)\cos\delta$, $\phi(x,\,t) = \chi(x,\,t)\sin\delta$, by replacing α_1 by $2\alpha_1$.

The initial conditions play an important role in the investigation of soliton interaction described by CNLSE. It has been shown in different works of the present authors [23, 92, 99, 101] that linearly polarized initial conditions, after the interaction, invariably lead to solitons for which both $\psi, \phi \neq 0$, and have in general elliptic polarizations. This means that a numerical solution for the general case is needed in order to provide initial conditions with elliptic polarization, which is the goal of the present work. A numerical solution for elliptic polarized stationary propagating solitons has been provided in [95]. In order to be able to use a numerically computed initial condition, one must have a code that gives the solution. Here, we propose a different numerical method to find the solution to be used in our future work on the interaction of solitons. Simultaneously, we can verify both our solution and the solution of [95].

The elliptically polarized solution has the form

$$\psi(x,\,t),\, \phi(x,\,t) = A_{\psi,\,\phi}(x - X - c_{\psi,\,\phi}t)$$

$$\times \exp\left\{ i \left[n_{\psi,\,\phi} - \frac{c_{\psi,\,\phi}}{2}(x - X - c_{\psi,\,\phi}t) + \delta_{\psi,\,\phi} \right] \right\}$$

(2.36)

where n_ψ and n_ϕ are the carrier frequencies and δ_ψ and δ_ϕ are the phases for the two components. Generally $n_\psi \neq n_\phi$, while $n_\psi = n_\phi$ only for circular polarization. The general polarization angle is defined as $\theta = \arctan(\max\{A_\phi\} / \max\{A_\psi\})$.

After some straightforward manipulations, we get the following system of conjugated real ordinary differential equations for the real-valued amplitudes A_ψ, A_ϕ:

$$\vec{F}(\vec{A}) \equiv \begin{pmatrix} A_\psi'' + \left(n_\psi + \frac{1}{4}c_\psi^2\right)A_\psi + \left[\alpha_1 A_\psi^2 + (\alpha_1 + 2\alpha_2)A_\phi^2\right]A_\psi \\ A_\phi'' + \left(n_\phi + \frac{1}{4}c_\phi^2\right)A_\phi + \left[\alpha_1 A_\phi^2 + (\alpha_1 + 2\alpha_2)A_\psi^2\right]A_\phi \end{pmatrix} = \begin{pmatrix} 0 \\ 0 \end{pmatrix}, \qquad (2.37)$$

where $\vec{A} = [A_\psi(x - X - c_\psi t), A_\phi(x - X - c_\phi t)]^T$ is the unknown vector function.

The above system has to be solved with the asymptotic boundary conditions

$$\lim_{x \to \pm\infty} A'_{\psi,\phi}(x - X - c_{\psi,\phi}t) = \lim_{x \to \pm\infty} A_{\psi,\phi}(x - X - c_{\psi,\phi}t) = 0. \qquad (2.38)$$

Thus, we are faced with solving the bifurcation boundary-value problem (BVP) (2.37) and (2.38) for the possible nontrivial solution. Introducing a fictitious time $\tau \geqslant 0$, lets us assume that the relationship holds

$$\frac{\partial \vec{F}}{\partial \vec{A}}\vec{\alpha} + \frac{\partial \vec{F}}{\partial \vec{A}''}\vec{\alpha}'' + \vec{F} = \vec{0}, \quad \text{with} \quad \vec{\alpha} \equiv \frac{\partial \vec{A}}{\partial \tau}. \qquad (2.39)$$

It is proved in [27] that the iteration

$$\vec{A}^{k+1} = \vec{A}^k + \Delta\tau_k \vec{\alpha}^k, \quad \Delta\tau_k = \frac{\|\vec{F}^k\|}{\|\vec{F}^k\| + \|\vec{F}^{k+1}\|}, \quad k = 0, 1, 2, \ldots$$

is quadratically convergent in the vicinity of the solution of (2.37) and (2.38). In the last formula, $\| \cdot \|$ stands for the L_2-norm. It is convenient to use cubic Hermitean splines on the uniform mesh

$$x_i = -x_\infty + ih, \qquad h = \frac{2x_\infty}{N}, \qquad i = 0, \ldots, N,$$

and the Gaussian points $1/2 \pm \sqrt{3}/6$ in [0,1] to be the collocation nodes.

We use the sech-like function as an initial approximation, i.e. $A_{\psi,\phi}^{(0)} \propto \operatorname{sech}(b_{\psi,\phi}x)$ and solve the above auxiliary BVP for given $n_{\psi,\phi}$, $c_{\psi,\phi}$, α_1 and α_2 with a spatial step $h = 10^{-2}$ and 'actual' infinity $x_\infty = 15 - 20$. Taking seven to eight iterations proves to be sufficient to reach the approximate nontrivial solution with L_2-norm of an error less than 10^{-12}.

In figure 2.26, we present three cases of elliptic polarization when the carrier frequency of the ψ-component is fixed ($n_\psi = -0.5$) and the carrier frequency of the ϕ-components $n_\phi \in [-0.7, -0.39]$. We were able to generate solutions with all possible angles of polarization in $\theta \in [0°, 90°]$ including the cases of circular and linear polarization. In this way, we found out that the elliptically polarized solitons do exist even for those values of the carrier frequency and phase velocity for which the linearly polarized sech-solutions do not exist. Note that the latter are limited by the condition $n_{\psi,\phi} + c_{\psi,\phi}^2/4 > 0$. Figure 2.26 shows that by varying the carrier frequency, one can produce solitons that have different supports. We mention here that as far as it can be judged by the graph, our shapes are in good quantitative agreement with the solution of [95].

Another dimension of complexity is introduced by the phases $\delta_{\psi,\phi}$ of the different components (see equation (2.36)), although these do not affect the amplitudes $A_{\psi,\phi}$. As shown in [101], the initial difference in phases can have a profound influence on the polarizations of the solitons after the interaction. In figure 2.27, we present a case with given amplitudes of envelopes but with different phases. Naturally, we present both the real and imaginary parts, because their relative shift is what matters in this case.

In the end, we present in figure 2.28 the interaction of two initial solitons with different polarizations. The left one has a circular polarization, while the right one

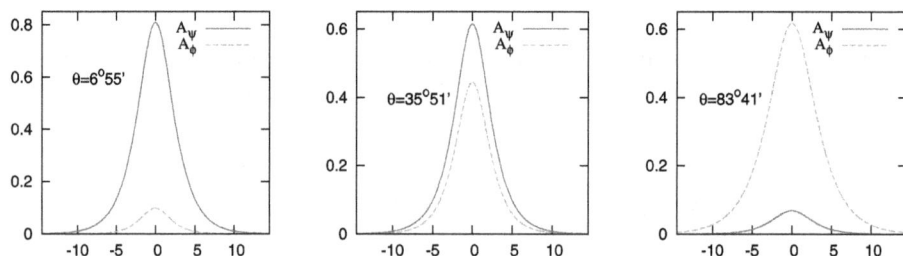

Figure 2.26. Amplitudes A_ψ and A_ϕ for $c_1 = -c_r = 1$, $\alpha_1 = 0.75$, $\alpha_2 = 0.2$, $n_\psi = -0.5$. Left: $n_\phi = -0.68$; middle: $n_\phi = -0.55$; right: $n_\phi = -0.395$.

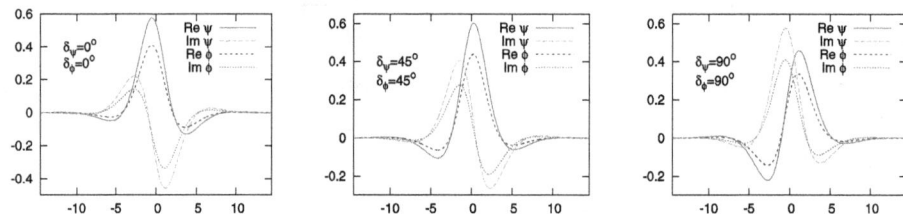

Figure 2.27. Real and imaginary parts of the amplitudes, $\Re\psi$, $\Im\psi$, $\Re\phi$, $\Im\phi$, from the case shown in the middle panel of figure 2.26 and the dependence on phase angle.

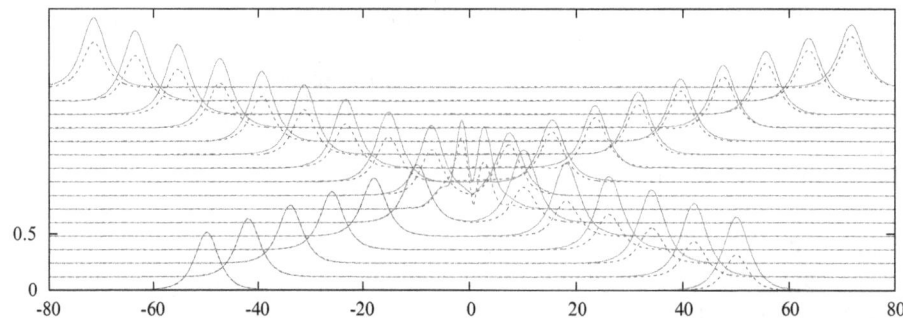

Figure 2.28. Collision dynamics of quasi-particles or $c_1 = -c_r = 1$, $\alpha_1 = 0.75$, $\alpha_2 = 0.2$. Left: circular polarization $n_{1\psi} = n_{1\phi} = -0.5$. Right: elliptic polarization $n_{r\psi} = -0.55$, $n_{r\phi} = -0.46$.

has elliptic polarization. The above mentioned exchange of polarizations is clearly seen. The stability of the computations are ensured by the conservative scheme we employ, but the lack of oscillations and dispersion testifies that the initial conditions are with high accuracy compatible with the equations.

We are already experienced to start to study the dynamics of elliptically polarized solitons. We combine various initial polarizations with various initial phase differences [105]. In particular, we establish that even for the Manakov system (trivial cross-modulation parameter $\alpha_2 = 0$), the polarization is changed after a head-on collision, depending on the initial phases of the solitons. For nontrivial α_2, a discontinuity of the individual polarization angles on the place of interactions ('polarization shock') is present as well. In more cases, the solitons survive as QPs with a changed polarization and insignificant phase-velocity shifts. In other cases of the initial phase difference, however, one of the solitons disappears after the head-on collision and becomes an energetic excitation or we observe newly born solitons (for large magnitudes of the cross modulation).

2.7 Collision dynamics of elliptically polarized solitons in vector nonlinear Schrödinger equation

We investigate numerically the collision dynamics of elliptically polarized solitons of the system of the CNLSE (VNLSE) for various different initial polarizations and phases. General initial elliptic polarizations (not *sech*-shape) include as particular cases the circular and linear polarizations. The elliptically polarized solitons are computed by a separate numerical algorithm described in the previous section. We find that, depending on the initial phases of the solitons, the polarizations of the system of solitons after the collision change, even for trivial cross-modulation. This sets the limits of the practical validity of the celebrated Manakov solution.

VNLSE is a soliton supporting system. Apart from its splendid performance in modeling the propagation of light pulses in fiber optics, it also offers the opportunity to investigate the quasi-particle behavior soliton. The latter is indispensable for the understanding of the fundamental phenomena associated with propagation of nonlinear waves. There are many different solitons supporting systems whose solutions behave as QPs, but NLSE and VNLSE exhibit the richest behavior and serve as very important testing grounds for the quasi-particle approach. For this reason, the VNLSE attracted the attention of leading researchers, and a number of excellent analytical results have been obtained over the years. For the integrable Manakov case (see [4, 5, 18, 26, 60, 72, 74, 86], among others), linear and circular polarization were treated in [69], and general polarization in [55, 112], among others. The initial polarization of VNLSE and its evolution during the quasi-particle interaction is a very important element, which is uniquely associated with the vector nature of the model. There are numerous experimental observations (see [24] and literature cited therein) of the effects connected with the polarization.

The existence of analytical solutions is usually limited to simpler models, and for the full fledged VNLSE with nontrivial cross-modulation, no such solution is available. In addition, an analytical solution cannot answer the questions related to its physical significance before its stability or robustness (in some general sense) is established. This can be done, e.g. via adequately devised numerical scheme. For this reason, the numerical schemes for VNLSE have a very important role to play. In order to adjust this scheme to VNSLE, internal iterations were applied in [23]. This concept yielded both a fully implicit scheme and implementation of the conservation laws on the difference level within the round-off error of the calculations. The above scheme was extended to complex arithmetic in [99] where the computer code for Gaussian elimination with pivoting of [21] was generalized for complex-valued multi-diagonal band algebraic systems.

Keeping in mind the importance of the initial polarization, we investigated in [101] the collision dynamics for circularly polarized solitons based on *sech*-functions and found out that there exists an infinite number of Manakov two-soliton solutions preceded by polarization discontinuity (shock) on the place of interaction. An interesting new result was that Manakov solitons with 45° initial polarization can emerge unchanged from the collision even for non-trivial cross-modulation, provided that the initial phases of both QPs are equal to zero. In order to enrich the range of investigations in the case of general polarization, we established an auxiliary conjugate system of nonlinear ordinary differential equations [102] in order to numerically generate initial elliptically polarized soliton solutions. We aim to conduct a series of simulations and to track the particle-like behavior of the interacted localized waves with arbitrary (elliptic) initial polarization.

Problem formulation

The system of CNLSE (VNLSE) in standard notations the system reads like (2.2). When $\alpha_2 = 0$, no nonlinear coupling is present despite the fact that 'cross-terms' proportional to α_1 appear in the equations [72, 114, 115]. In this particular case $\psi = \cos(\theta)\chi$, $\phi = \sin(\theta)\chi$, where χ is the solution of a single NLSE with nonlinearity coefficient $\sqrt{2}\,\alpha_1$.

Here, we concern ourselves with the soliton solutions which are localized envelopes on a propagating carrier wave. This defines the type of the initial conditions to be used. The general form of the latter is

$$\chi(x, t; X, c, n_\chi) = A_\chi(x + X - ct) \exp\left\{ i\left[n_\chi t - \frac{1}{2}c(x - X - ct) + \delta_\chi \right] \right\} \quad (2.40)$$

where χ stands for either ψ or ϕ, c is the phase speed of the envelope, X is the initial position of the center of the soliton, $\vec{n}\,(n_\psi, n_\phi)$ is the vector of carrier frequencies of the components, and $\vec{\delta}\,(\delta_\psi, \delta_\phi)$ is the phase vector of the two components. Note that the phase speed is the same for the two components ψ and ϕ. If they propagate with different phase speeds, after some time, the two components will be in two different positions in space, and will no longer form a single structure. At the same time, the carrier frequencies of the different components can be different, i.e. $n_\psi \neq n_\phi$. In such a case, one is faced with the so-called elliptically polarized solitons of CNLSE, and the two components of the envelope are governed by the coupled system of differential equations (2.37). This formulation includes the particular case of a circular polarization, when carrier frequencies are equal to each other, i.e. $n_\psi = n_\phi$. Then, for the envelopes, one has $A_\psi = A(x)\cos(\theta)$, $A_\phi = A(x)\sin(\theta)$, ($\theta \equiv \arctan(\max|\phi|/\max|\psi|)$). When $\theta = 0°$ or $\theta = 90°$, the circular polarization reduces to the so-called linear polarization, in which one of the components is identically equal to zero.

Obviously, the trivial solution of equation (2.37) always persists for asymptotic boundary conditions, and the nontrivial (if it exists) is the result of a bifurcation. The problems associated with the bifurcation are elucidated in [95]. We provided another numerical solution to equation (2.37) in [102], which can be readily used as an initial condition for the unsteady computations.

Before turning to the numerical investigation, we remind you here that the system equation (2.2) possesses at most three conservation laws when asymptotic boundary conditions are imposed, which means that for this kind of boundary condition it is non-fully integrable.

Results and discussion

The phenomenology of the CNLSEs is very rich, and the interaction of the solitons is strongly influenced by all different parameters: nonlinearity, cross-modulation, carrier frequencies, phase speed, initial polarization of the solitons. We have set the goal to better understand the polarization dynamics.

Manakov solitons ($\theta = 45°$). The role of the phase
We begin with the case when there is no cross-modulation, i.e. $\alpha_2 = 0$. This case is alternatively known as the Gross–Pitaevskii or Manakov system (MS). The solution was found by Manakov [72] through reducing the system to a single NLSE by setting $\psi = \phi$. In our notations, this corresponds to the case $\theta = 45°$. Actually, the fact that $\alpha_2 = 0$ allows the Manakov solution to exist for any θ, and the shape functions of the envelopes are the same *sech*-es. Yet, the system in this case is still nonlinear, and it is very important to investigate numerically the collisions of the *sech*-solitons with the goal to examine if the Manakov solution is robust and if it survives the interaction. To this end, we performed several numerical experiments with different initial phase speeds of the solitons, and various phases. Changing the phase speed c, does not seem to lead to qualitative differences of the dynamics, and we present here the results for one representative case $c_1 = -c_r = 1$ (the subscripts '*l*' and '*r*' refer to the left and right soliton, respectively). Similarly, the effect of α_1 is straightforward, and we will discuss the case of $\alpha_1 = 0.75$. We select the two components to have the same carrier frequency $n_{l\psi} = n_{r\psi} = n_{l\phi} = n_{r\phi} = -1.5$ and focus on the effect of the initial phases. We treat two different cases: (a) $\delta_l = \delta_r = 0°$ and (b) $\delta_l = 0°$, $\delta_r = 45°$. The results are presented in figure 2.29. The most important observation is that the Manakov solution is realized only when the initial phases are equal to each other. When one of the phases differs from the other by 45°, the two initially circularly polarized solitons emerge after the interaction as elliptically polarized solitons with $\theta = 36° \ 17'$ and $\theta = 53° \ 37'$, respectively. This result is very

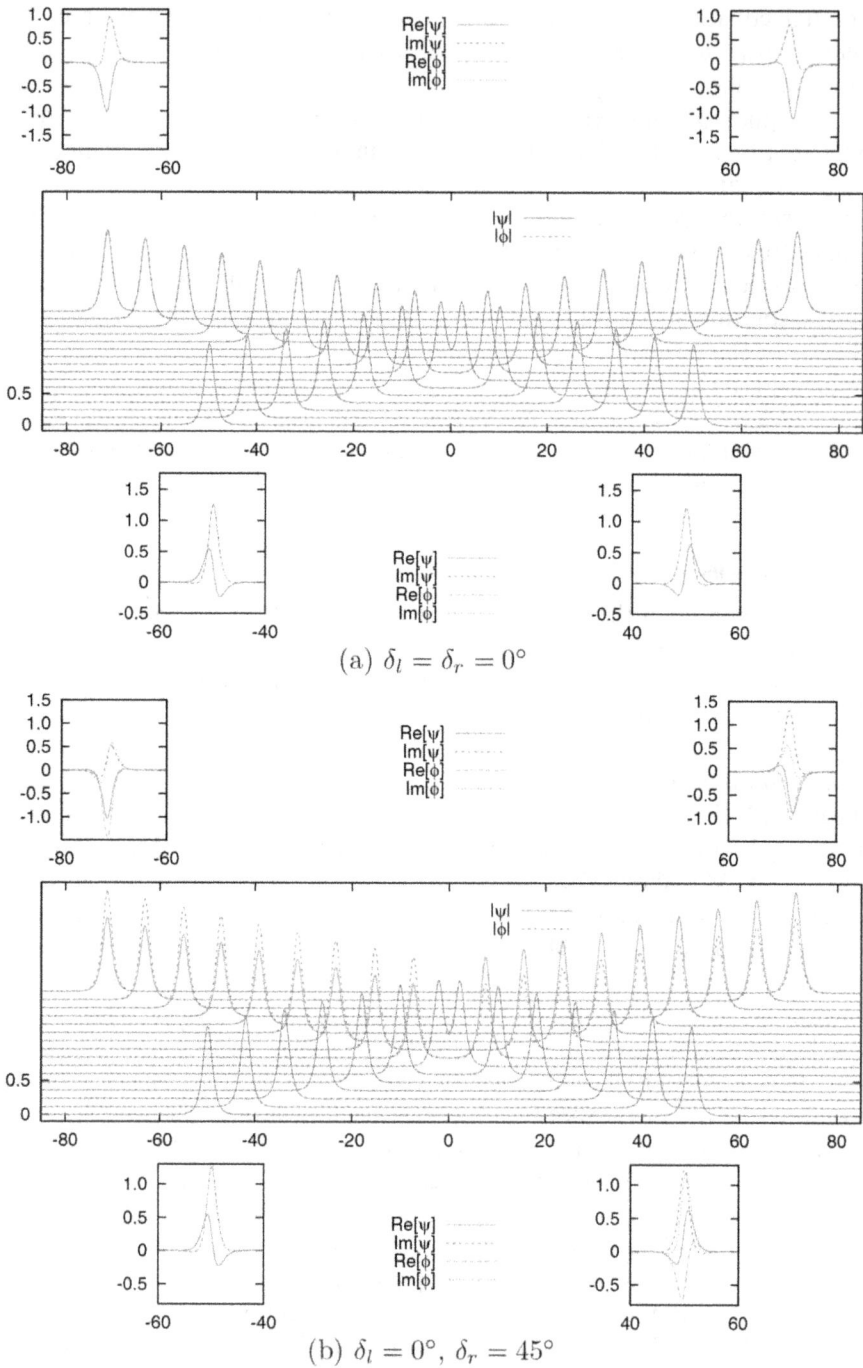

Figure 2.29. The role of the initial phases on the interaction of initial circularly polarized solitons with $\theta = 45°$ for $\alpha_2 = 0$, $\alpha_1 = 0.75$, $c_1 = -c_r = 1$, $n_{l\psi} = n_{r\psi} = n_{l\phi} = n_{r\phi} = -1.5$.

significant, and it points out towards a bifurcation. The solution presented in figure 2.29 coexists with the Manakov solution (the latter is valid for any value of the initial phases). Apparently, the solution obtained here is robust, while Manakov is not.

The 'fragility' of Manakov solitons was first addressed in [101], where an in-depth study was conducted for various initial configurations for the soliton parameters. The summary of those results is presented in table 2.5. We have found as well that the phases of the components play an essential role when the system 'chooses' one or another polarization angle of the one-soliton solutions that break away from the site of collision. As a rule, the initial polarization is not retained by the solution and outside the cross-section of the interaction, a discontinuity is observed of the polarization which can be called a 'polarization shock' and it takes place during the interaction despite the fact that the ingoing and outgoing functions are smooth. For nonzero phase, one can see that after the interaction, the two QPs become different Manakov solitons than the original two that entered the collision. Apparently, the specific choice of the outgoing polarization angle is controlled by the magnitude of phase difference $[\![\delta]\!] \equiv \delta_r - \delta_l$. The comparison with the relevant case (with analytic initial condition) considered in [101] enables us to conclude that the envelopes in this case are *sech*-like functions.

In the cited work [101], an important fact of the interaction of polarized vector solitons was unearthed, namely the approximate law of conservation of the total polarization of the system of solitons. Since the conservation was beyond doubt for initially circularly polarized solitons, it seemed important to investigate the issue deeper, and to investigate the polarization dynamics when the initial polarization is not restricted to the circular case.

Now we can move to the case of a non-trivial cross-modulation parameter. A very important observation was made in [101], namely that a Manakov type of solution can also be found analytically for non-trivial cross-modulations, but only for one single value of the polarization: $\theta = 45°$. The question arises of what kind of evolution this soliton undergoes during the interaction with another soliton of the same type. For definiteness, we choose $\alpha_2 = 4$. Figure 2.30 shows the result for this case. In order to test and validate the algorithm for computing the initial condition, here we use the numerical solution rather than the analytical *sech*-solution for A_ψ and A_ϕ. The results are indistinguishable from those obtained with the analytical initial condition. It is interesting to mention here that the interaction is relatively robust with respect to changes of the initial phase in the interval $0 \leqslant \delta \leqslant 50°$. Outside of that interval, the impact of the initial phase is very strong. Not just the polarization suffers a jump (as already observed in the plain Manakov case), but the phase speeds of the outgoing solitons are radically changed after the interaction. The sensitivity to the values of the initial phases is very strong in the interval $[\![\delta]\!] \in [90°, 180°]$. As can be seen in the figures 2.30(c) and (d), the change of the initial polarization difference from 90° to 135° changes the outgoing configuration radically. While for $[\![\delta]\!] = 90°$, the left going soliton acquires a larger phase speed, in the case $[\![\delta]\!] = 135°$, it is the right-going soliton that is sped up significantly. Respectively, their counterparts are slowed down. This effect is in accordance with the laws of conservation of mass and momentum.

Table 2.5. Polarization evolution during collisions.

(a) Initial values

$[[\delta]]$	θ_1^i	θ_r^i	$\theta_1^i + \theta_1^r$	m_ψ^i	θ_ϕ^i	p^i	e^i	α_2
$0°$	$45°$	$45°$	$90°$	2.9837653	2.9790827	$-0.11139648 \times 10^{-10}$	-2.0016750	0
$45°$	$45°$	$45°$	$90°$	2.9837653	2.9790827	$-0.11139716 \times 10^{-10}$	-2.0016750	0
$0°$	$45°$	$45°$	$90°$	0.46976218	0.46976218	$0.24620106 \times 10^{-10}$	-0.31538994	4
$45°$	$45°$	$45°$	$90°$	0.46976218	0.46976218	$0.24620078 \times 10^{-10}$	-0.31538994	4
$270°$	$45°$	$45°$	$90°$	0.46976218	0.46976218	$0.24619929 \times 10^{-10}$	-0.31538994	4
$0°$	$50° \, 08'$	$50° \, 08'$	$100° \, 16'$	0.58533758	0.92145185	$0.61555635 \times 10^{-11}$	-0.25915818	2
$45°$	$50° \, 08'$	$50° \, 08'$	$100° \, 16'$	0.58533758	0.92145185	$0.61556880 \times 10^{-11}$	-0.25915818	2
$90°$	$50° \, 08'$	$50° \, 08'$	$100° \, 16'$	0.58533758	0.92145185	$0.61556301 \times 10^{-11}$	-0.25915818	2
$135°$	$50° \, 08'$	$50° \, 08'$	$100° \, 16'$	0.58533758	0.92145185	$0.61557896 \times 10^{-11}$	-0.25915818	2
$140°$	$50° \, 08'$	$50° \, 08'$	$100° \, 16'$	0.58533758	0.92145185	$0.61557896 \times 10^{-11}$	-0.25915818	2
$160°$	$50° \, 08'$	$50° \, 08'$	$100° \, 16'$	0.58533758	0.92145185	$0.61557427 \times 10^{-11}$	-0.25915818	2
$180°$	$50° \, 08'$	$50° \, 08'$	$100° \, 16'$	0.58533758	0.92145185	$0.61556354 \times 10^{-11}$	-0.25915818	2
$270°$	$50° \, 08'$	$50° \, 08'$	$100° \, 16'$	0.58533758	0.92145185	$0.61556354 \times 10^{-11}$	-0.25915818	2
$305°$	$50° \, 08'$	$50° \, 08'$	$100° \, 16'$	0.58533758	0.92145185	$0.61556354 \times 10^{-11}$	-0.25915818	2
$0°$	$39° \, 51'$	$50° \, 08'$	$89° \, 59'$	0.75339470	0.75339471	$-0.26425613 \times 10^{-07}$	-0.25915813	2
$45°$	$39° \, 51'$	$50° \, 08'$	$89° \, 59'$	0.75339470	0.75339471	$-0.26425613 \times 10^{-07}$	-0.25915813	2
$135°$	$39° \, 51$	$50° \, 08$	$89° \, 59'$	0.75339470	0.75339471	$-0.26425613 \times 10^{-07}$	-0.25915813	2
$0°$	$45°$	$50° \, 08'$	$95° \, 08'$	0.69922661	0.86728375	$0.59265539 \times 10^{-01}$	-0.40253479	2
$45°$	$45°$	$50° \, 08'$	$95° \, 08'$	0.69922661	0.86728375	$0.59265539 \times 10^{-01}$	-0.40253479	2
$135°$	$45°$	$50° \, 08'$	$95° \, 08'$	0.69922661	0.86728375	$0.59265539 \times 10^{-01}$	-0.40253479	2

(*Continued*)

Table 2.5. (Cont.) Values after the interaction.

$[[\delta]]$	θ_l^f	θ_r^f	$\theta_l^f + \theta_r^f$	m_ψ^f	m_ϕ^f	p^f	e^f	α_2
0°	45°	45°	90°	2.9837653	2.9790827	$-0.14031559 \times 10^{-02}$	-2.0016748	0
45°	36° 17'	53° 37'	89° 54'	2.9837653	2.9790827	$0.35851357 \times 10^{-02}$	-2.0016748	0
0°	45°	45°	90°	0.4697618	0.46976218	$0.66569146 \times 10^{-03}$	-0.31538991	4
45°	42° 38'	47° 49'	90° 27'	0.4697618	0.46976218	$0.83149016 \times 10^{-03}$	-0.31538097	4
270°	47° 50'	43° 09'	90° 59'	0.4697618	0.46976218	$-0.69621695 \times 10^{-03}$	-0.31535741	4
0°	48° 59'	49° 05'	98° 04'	0.58533758	0.92145185	$0.14500572 \times 10^{-04}$	-0.25895854	2
45°	49° 07'	49° 17'	98° 14'	0.58533758	0.92145185	$0.19469770 \times 10^{-04}$	-0.25902622	2
90°	55° 19'	45° 41'	101° 0'	0.58533758	0.92145185	$0.50089322 \times 10^{-03}$	-0.25910314	2
135°	58° 45'	44° 56'	103° 41'	0.58533758	0.92145185	$-0.21591801 \times 10^{-03}$	-0.25913752	2
140°	59° 57'	44° 54'	104° 51'	0.58533758	0.92145185	$-0.87138012 \times 10^{-04}$	-0.25914128	2
160°	39° 45'	57° 37'	97° 52'	0.58533758	0.92145185	$0.51123279 \times 10^{-03}$	-0.25915436	2
180°	51° 25'	48° 25'	99° 50'	0.58533758	0.92145185	$-0.67917481 \times 10^{-04}$	-0.25915805	2
270°	44° 50'	56° 30'	101° 20'	0.58533758	0.92145185	$-0.42974918 \times 10^{-03}$	-0.25912060	2
305°	48° 54'	50° 32'	99° 26'	0.58533758	0.92145185	$-0.46829621 \times 10^{-03}$	-0.25907364	2
0°	41° 37'	50° 42'	92° 29'	0.75339470	0.75339471	$-0.46310191 \times 10^{-04}$	-0.25915676	2
45°	37° 49'	52° 11'	90°	0.75339470	0.75339471	$0.68540316 \times 10^{-03}$	-0.25915751	2
135°	46° 32'	43° 23'	89° 58'	0.75339470	0.75339471	$0.70073356 \times 10^{-03}$	-0.25915432	2
0°	42° 11'	52° 53'	95° 04'	0.69922661	0.86728375	$0.58914178 \times 10^{-01}$	-0.40253371	2
45°	45° 04'	49° 40'	94° 44'	0.69922661	0.86728375	$0.60009911 \times 10^{-01}$	-0.40253265	2
135°	48° 47'	46° 2'	94° 49'	0.69922661	0.86728375	$0.59821246 \times 10^{-01}$	-0.40253466	2

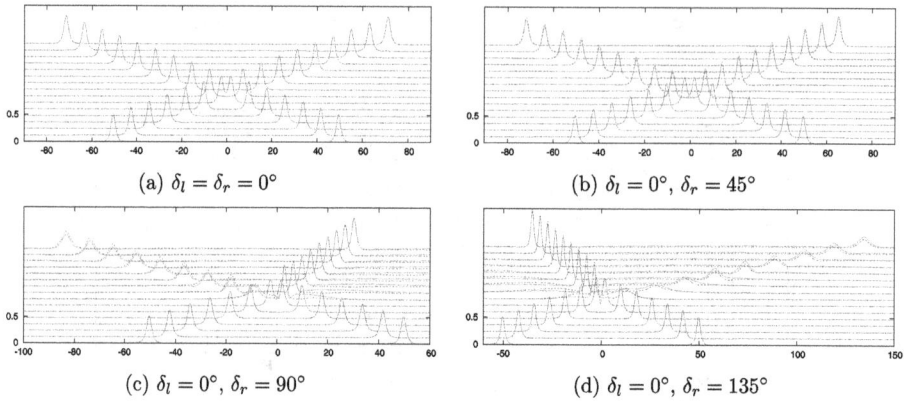

(a) $\delta_l = \delta_r = 0°$

(b) $\delta_l = 0°$, $\delta_r = 45°$

(c) $\delta_l = 0°$, $\delta_r = 90°$

(d) $\delta_l = 0°$, $\delta_r = 135°$

Figure 2.30. Initial circular polarization $\theta_1 = \theta_r = 45°$ for $\alpha_2 = 4$.

Cases with elliptically polarized solitons as initial conditions

We chose $\alpha_2 = 2$ and settle on the following equal initial polarizations $\theta_1 = \theta_r = 50°$ 08', which are the results of the different carrier frequencies of the different components of the initial solitons, namely $n_{1\psi} = n_{r\psi} = -1.5$, $n_{1\phi} = n_{r\phi} = -1.1$, $c_1 = -c_r = 1$, $c_l = -c_r = 1$, $\alpha_1 = 0.75$, $\alpha_2 = 2$ and focus again on the effect $[\![\delta]\!]$. In figures 2.31, the results of the interaction for different initial phases are presented. One can see that for $[\![\delta]\!] < 90°$, the interaction is virtually elastic, and even the polarization of each soliton is barely changed. When $[\![\delta]\!] > 90°$, the result changes dramatically: the polarization of the left going soliton is larger than the polarization of the right outgoing soliton. Once again, a bifurcation takes place and after the interaction the polarizations are appreciably different from the initial one. Because of the symmetry of the problem, one should expect to encounter both kind of patterns presented in figures 2.31 and their mirror images (horizontally flipped). The choice of the particular configuration is made by the round-off properties of the algorithm, as it is usually the case with bifurcations. In order to make sure that there is not some persistent small inaccuracy of the algorithm that leads to the selection, we took the results for the last two time stages, flipped them horizontally and plugged them in the difference scheme. We found that the residue was of the order 10^{-6} which is much better than the truncation error of the scheme. This proves that no approximation error is present in the scheme, even a very insignificant one. In case the residue is of order of the truncation error, this might mean that the truncation error of the implemented algorithm differs from the theoretical truncation error of the scheme. Clearly, this is not the case in the presented algorithm.

A very interesting and rather nonintuitive effect is observed for $[\![\delta]\!] \in [130°, 140°]$ in figures 2.31(e) and (f): one of the QPs loses a significant portion of its 'mass' (as testified by the decreased amplitude), contributing it to the other QP. It still carries an appreciable momentum because of its increased phase speed. The other outgoing soliton has a larger mass but a smaller phase speed. This effect is not exactly what is called 'trapping', because the momentum and the energy of one of the solitons is not entirely transferred to the other, but it seems akin to trapping as

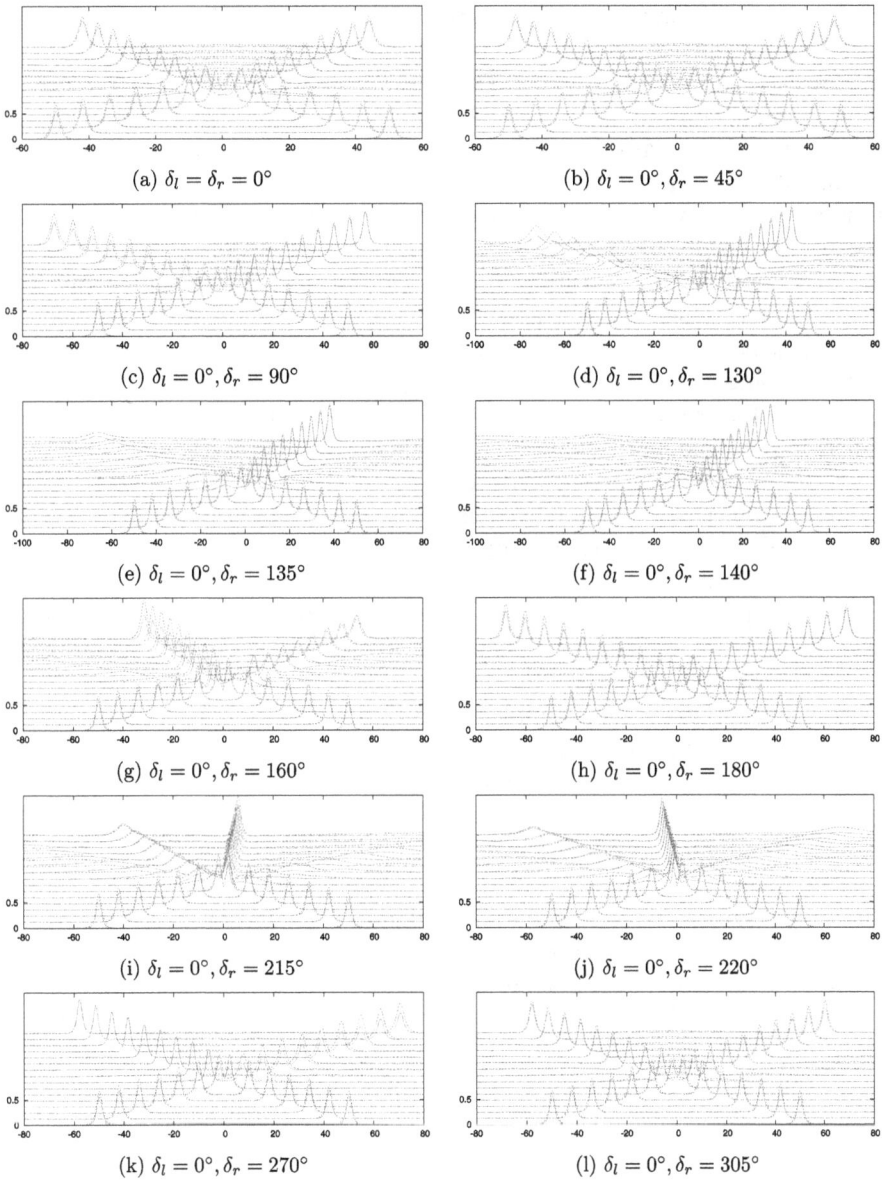

(a) $\delta_l = \delta_r = 0°$

(b) $\delta_l = 0°, \delta_r = 45°$

(c) $\delta_l = 0°, \delta_r = 90°$

(d) $\delta_l = 0°, \delta_r = 130°$

(e) $\delta_l = 0°, \delta_r = 135°$

(f) $\delta_l = 0°, \delta_r = 140°$

(g) $\delta_l = 0°, \delta_r = 160°$

(h) $\delta_l = 0°, \delta_r = 180°$

(i) $\delta_l = 0°, \delta_r = 215°$

(j) $\delta_l = 0°, \delta_r = 220°$

(k) $\delta_l = 0°, \delta_r = 270°$

(l) $\delta_l = 0°, \delta_r = 305°$

Figure 2.31. $\theta_1 = \theta_r = 50° \, 08'$, $\alpha_2 = 2$.

far as the diminished amplitude of one of the solitons is concerned. It goes beyond the scope of the present work to dwell deeper on the 'trapping' for the system under consideration, and it will be done elsewhere.

For $[\![\delta]\!] \geqslant 160°$, the outcome of the interaction is radically different. Now, the left going soliton has a large amplitude and is relatively slower (note that the right-going soliton has a slower phase speed than the initial phase speed). Now the polarization of the left-going soliton is also significantly larger than for the other cases. Increasing further $[\![\delta]\!]$ to 180° does not qualitatively change the interaction, but merely mitigates the effect of the changes. The intuitive expectation that $[\![\delta]\!]$ = 180° may have the same quantitative effect as $[[\delta]]$ = 0° is not confirmed. Clearly, if the two components of the initial vector solitons are in anti-phase, it is not equivalent to the case when they are in phase ($[\![\delta]\!]$ = 0°).

For this reason, we also investigated the interaction in the interval $[\![\delta]\!] \in$ (180°, 360°), and the results are presented in the respective panels of figure 2.31. We have discovered another very sensitive interval $[\![\delta]\!] \in$ (210°, 230°), where one of the solitons becomes much steeper, and a third very fast soliton with relatively small amplitude is born whose momentum balances the momentum of the two large solitons escaping to the left (see figures 2.31(j) and (i) to a lesser extent). The rest of the values 270 < $[\![\delta]\!]$ < 360° do not lead to such violent changes, and the effect is confined to some moderate changes of the soliton polarizations.

Similarly to the circularly polarized solitons, the sum of polarizations of the outgoing solitons is approximately equal to the sum of the original ones. Thus, one can once again argue the case for 'conservation' of the polarization. The numerical values are presented in table 2.5.

The next experiment is to investigate the case where the two initial solitons have elliptic polarization and these polarizations are complementary to each other. We chose one of the solitons to have the same parameters as the previous case, and take the other one to have the 'complementary' polarization. To construct such a configuration, we take $n_{l\psi} = n_{r\phi} = -1.1$, $n_{l\phi} = n_{r\psi} = -1.5$. We keep all the other parameters the same: $c_1 = -c_r = 1$, $\alpha_1 = 0.75$, $\alpha_2 = 2$. For the initial polarization, we obtain: $\theta_1 = 39° \ 51'$, $\theta_r = 50° \ 08'$, $\theta_1^\circ + \theta_r^\circ \approx 90°$.

The results are presented in figure 2.32 for four different values of $[\![\delta]\!]$. The results can be summarized as follows: for $[\![\delta]\!]$ = 0° (figure 2.32(a)), the polarizations are virtually the same as for the initial configuration, with a slight tendency to diminishing the difference between them (see table 2.5). For $[\![\delta]\!]$ = 45° (figure 2.32(b)), the polarizations are close to the initial configuration, with a slight tendency to increasing the difference between them (see table 2.5), which is qualitatively

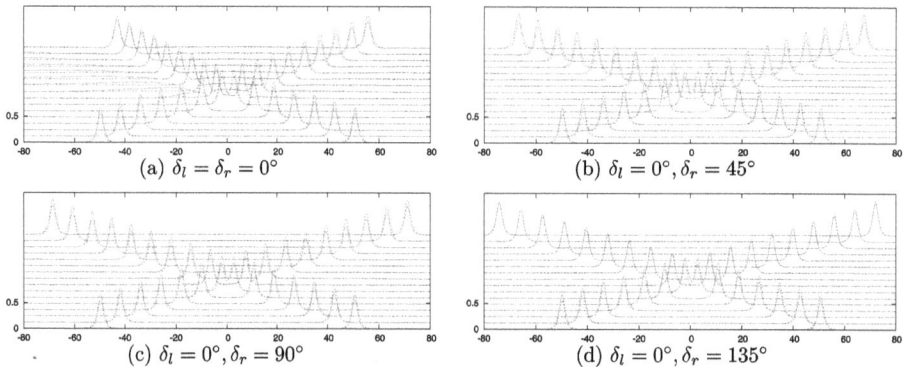

Figure 2.32. $n_{l\psi} = n_{r\phi} = -1.1$, $n_{l\phi} = n_{r\psi} = -1.5$ ($\theta_1 = 39° \ 51'$, $\theta_r = 50° \ 08'$) for $\alpha_2 = 2$.

different from the case $[\![\delta]\!] = 0°$. There is not much difference between the cases $[\![\delta]\!] = 90°$ (figure 2.32(c)) and $[\![\delta]\!] = 45°$ (figure 2.32(b)). Finally, for $[\![\delta]\!] = 135°$ (figure 2.32(d)), the tendency is reversed back to diminishing the difference between the outgoing polarizations.

In all of the cases presented in figure 2.32, a 'slight' polarization shock is experienced after the interaction of the solitons, but once again an approximate conservation of the total polarization is observed (table 2.5).

The table also gives info on the 'masses' of the solitons, pseudomomentum, and total energy of the system of solitons. One can see the perfect conservation of the mass and the acceptable conservation of energy. The pseudomomentum is conserved within the truncation order of the scheme. For all the cases considered in the paper, the total pseudomomentum is expected to be equal to zero due to symmetry. It is worth noting here that the exact conservation of the pseudomomentum is to be expected only for asymptotic boundary conditions. Even when the computational box is large enough, there are some nontrivial values of the derivatives of the sought function near the boundaries where the functions themselves are kept equal to zero. These small values are of order of the truncation error, and they contribute to the small deviation of the computed pseudomomentum from zero.

Conclusions

The polarization dynamics of interacting vector solitons of the CNLSE (VNLSE) is investigated numerically by means of an energy conserving difference scheme. The initial condition is obtained numerically by solving the system of ODEs for the shape of the stationary propagating vector soliton. For the latter, the carrier frequencies of the different components of the vector soliton are different and the amplitude of the envelope is no longer a *sech*. Different combinations of elliptically polarized (e.g. circularly polarized for the particular case of angle of polarization 45°) solitons are used as the initial condition and the evolution of the system is followed numerically. Using a general initial polarization significantly enriches the observed effects of the nonlinear coupling in CNLSE. The evolution of the polarization during the collision of elliptically polarized solitons is the object of the investigation. Our results show that the polarization dynamics turns out to be very susceptible to the initial phases of the solitons, which appears to be a novel result, not known from the literature.

First, we have investigated the case of trivial cross-modulation, $\alpha_2 = 0$. For this case, the Manakov analytical solution is available. Our computations show that there exists an initial phase for which the Manakov solutions persists after the interaction, but there are also values of the phases for which the polarization changes to elliptic after the interaction. This means that a bifurcation takes place, and along with the solution with constant polarization (the analytical Manakov solution), another solution appears for which a kind of jump (or 'shock') of the polarization takes place during the cross-section of the interaction. This result is very important to put the Manakov solution in the proper perspective.

Second, we treat cases with nontrivial cross-modulation, $\alpha_2 \neq 0$. We found that the actual magnitude of α_2 has only a quantitative effect on the results, and the qualitative effect is present even for $\alpha_2 < 1$. Our results show that the initially circular polarization inevitably transforms to elliptic polarization of the outgoing solitons. What is more important is the fact that the nontrivial cross-modulation leads to an increase or decrease of the phase speed of the outgoing solitons, which also brings about drastic changes of their amplitudes. In some cases, the amplitude is so small that the result can be classified as 'trapping' in the sense that one of the outgoing solitons captures the predominant energy of the process, while the other appears more as a small disturbance. The effects scale with α_2, in the sense that for larger α_2, the mentioned deviations are bigger. Thus, one of the main results of the present paper is in identifying the role of the initial phase, showing that it can have a dramatic effect on the interaction for non-trivial cross-modulations.

In all considered cases, we find that although the initial polarizations can suffer a shock (or a jump) during the interaction, the total polarization is fairly well conserved. The actual polarization angles are strongly influenced by the initial phases, but their sum is virtually independent of the phase.

Our next goal is the linear coupling and its effect on the soliton dynamics [103]. We consider a superposition of two linearly polarized soliton envelopes and two kinds of interaction—head-on collision and taking over. In both cases, they behave like QPs after the interaction. When a nontrivial linear coupling is present, a new kind of polarization is observed—the rotational one. It is shown by so-called soliton breathing caused by a periodic mass exchange between the soliton components conserving the total mass of the envelopes. The rotational polarization does not depend on the interaction. Generally speaking, the magnitude of the linear coupling, coefficient Γ, is a complex number. Then, the rotational polarization is complemented with a blow-up or self-dissipation of the solitons.

2.8 Collision dynamics of polarized solitons in linearly coupled vector nonlinear Schrödinger equation

Investigating the soliton dynamics in both linearly and nonlinearly CNLSEs is of great importance from several different perspectives. The main ones are the propagation of optical pulses in optical fibers [9, 52] and the investigation of the QP behavior of soliton solutions. The essential new feature of CNLSEs in comparison with the single NLSE is the polarization, which is related to the relative amplitudes of the components. Keeping in mind the fact that each component is actually a complex-valued function, one can appreciate the complexity of the possible soliton interactions. The role of the nonlinearity in the interaction of initially linearly and elliptically polarized solitons was investigated in [99, 100]. It was uncovered that depending on the magnitude of the cross-modulation parameter (presenting the nonlinear coupling), the interaction between the modes during the collision changes the polarization of the QPs, and/or gives birth to one or more QPs. On the other hand, the CNLSE model has a richer phenomenology when a linear coupling is considered alongside with the main, nonlinear coupling (see, for example, [92] and the literature cited therein). This quantity generates rotational polarization, which is independent of the initial polarization of the soliton system.

This is the reason to focus our attention on the dynamics of the soliton solutions in the MS (cross-modulation $\alpha_2 = 0$) when a linear coupling is present.

Problem formulation

Let us consider a linearly coupled NLSEs (LCNLSEs)

$$i\psi_t = \beta\psi_{xx} + \alpha_1(|\psi|^2 + |\phi|^2)\psi + \Gamma\phi, \qquad (2.41a)$$

$$i\phi_t = \beta\phi_{xx} + \alpha_1(|\phi|^2 + \psi|^2)\phi + \Gamma\psi, \qquad (2.41b)$$

where β is the dispersion coefficient, α_1 parameterizes the self-focusing in birefringent media. This system possesses three conservation laws when asymptotic boundary conditions are imposed, namely when $\psi, \phi \to 0$ for $x \to \pm\infty$. Following [23, 92], we define 'mass', M, (pseudo)momentum, P, and energy, E as follows

$$M \stackrel{\text{def}}{=} \frac{1}{2\beta}\int_{-L_1}^{L_2}(|\psi|^2 + |\phi|^2)dx, \quad P \stackrel{\text{def}}{=} -\int_{-L_1}^{L_2}\Im(\psi\bar{\psi}_x + \phi\bar{\phi}_x)dx, \quad E \stackrel{\text{def}}{=} \int_{-L_1}^{L_2}\mathcal{H}dx, \quad (2.42)$$

where

$$\mathcal{H} \stackrel{\text{def}}{=} \beta(|\psi_x|^2 + |\phi_x|^2) - \frac{\alpha_1}{2}(|\psi|^2 + |\phi|^2)^2 + 2\Gamma[\Re(\bar{\psi}\phi)]$$

is the Hamiltonian density of the system. Here, $-L_1$ and L_2 are the left end and the right end of the interval under consideration. The following conservation/balance laws hold, namely

$$\frac{dM}{dt} = 0, \qquad \frac{dP}{dt} = \mathcal{H}|_{x=L_2} - \mathcal{H}|_{x=-L_1}, \qquad \frac{dE}{dt} = 0, \qquad (2.43)$$

which means that for asymptotic boundary conditions LCNLSEs admit at least three conservation laws. Unfortunately, there is no indication in the literature that the systems (2.41a) and (2.41b) admit more conservation laws or that they are fully integrable.

The system under consideration is of Manakov type with additional linear coupling that leads to oscillation of the maximal height of the localized pulses ('breathing'), even when they are not noninteracting. This can be shown analytically following [92] through the substitution

$$\psi = \Psi\cos(\Gamma t) + i\Phi\sin(\Gamma t), \qquad (2.44a)$$

$$\phi = \Phi\cos(\Gamma t) + i\Psi\sin(\Gamma t), \qquad (2.44b)$$

which reduces the original linearly coupled system equations (2.41) to the ubiquitous MS for functions Ψ and Φ. In order to create a numerical tool that allows expansion to more complicated cases, we will solve the original system (2.41) rather than the reduced system for Ψ and Φ.

The linear coupling parameter Γ can be, in general, a complex number. The real part $\Gamma_r = \Re[\Gamma]$ governs the oscillations between states termed as breathing solitons, while the imaginary part $\Gamma_i = \Im[\Gamma]$ is responsible for the gain/dissipation behavior of soliton solutions. Equations (2.41) possess solutions, which are combinations of interacting solitons oscillating with frequency Γ_r. These solutions are pulses whose modulation amplitude is of a general form (non-*sech*) and their polarization rotates with time. This determines the choice of initial conditions for numerical investigation of temporal evolution of interacting solitons.

The soliton solutions (QPs) are localized envelopes on a propagating carrier wave. For streamlining the notation, it is convenient to introduce the vectors $\vec{\chi} = (\psi, \phi)^T$ and $\vec{\delta} = (\delta_\psi, \delta_\phi)^T$. Then, the initial condition is constructed as the superposition of two QPs situated at X_l and X_r, and propagating with phase speeds c_l and c_r, i.e.

$$\vec{\chi}(x, 0) = \vec{\chi_l}(x, 0, c_l, n_l, X_l, \vec{\delta_l}) + \vec{\chi_r}(x, 0, c_r, n_r, X_r, \vec{\delta_r}). \tag{2.45}$$

Here, $|X_r - X_l|$ has to be large enough so the 'tail' of the first QP is fairly well decayed at the position of the second QP. In this paper, the scalar form of each QP in the right-hand side of (2.45) is assumed to be *sech*-like, i.e.

$$(\psi, \phi)^T(x, t; X, c, n) = (A_\psi, A_\phi)^T \, \text{sech}[b(x - X - ct)]$$
$$\times \exp\left\{ i\left[\frac{c}{2\beta}x + nt + \delta_{\psi,\phi} \right] \right\} \tag{2.46}$$

where n is the carrier frequency, and $\delta_{\psi,\phi}$ are the phases of the two components. Note that the phase speed is the same for the two components ψ and ϕ. If they propagate with different phase speeds, the two components will be in two different positions in space after some time, no longer forming a single structure.

Results and discussion

We observe that the time oscillation ('breathing') of the amplitude of the soliton does not interfere with the soliton collision, i.e. the breathing of the pulses take place even without any interaction. This distinguishes the solitons considered here from the breathers previously reported in the literature [20, 111]. One can say that the apparent breathing is actually a manifestation of the rotation of the polarization. In a sense, the linear coupling (parameterized by Γ) is responsible for the exchange of wave mass between the modes. Following [92], we call this effect 'cross-dispersion' of the signals. In figures 2.33 and 2.34, we present these features for given Γ in both cases—head-on and takeover interaction.

Our observations show that the presented here interactions are independent of the initial phase difference $[\![\delta]\!]$ and the trajectories of the centers of the QPs experience an almost insignificant shift after the interaction. Thus, in the case of head-on collision given in figure 2.33(a), initial trajectories are given by $x = \pm 45 \mp t$ while the trajectories of the outgoing solitons are given by $x = \pm 44.7 \mp t$ (figure 2.33(b)). The pseudomomentum varies between 10^{-15} (it vanishes due to the symmetry) in the

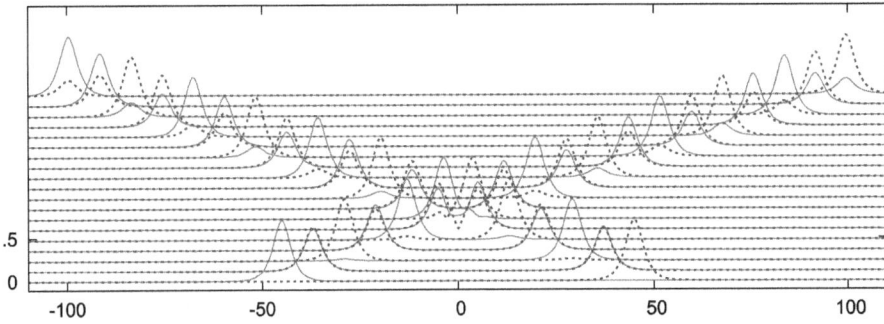

a) Profiles of the components

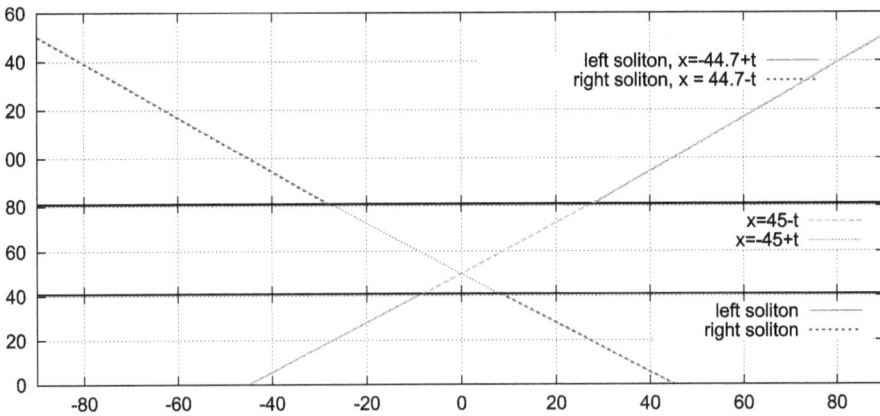

left soliton, x=-44.7+t
right soliton, x = 44.7-t

x=45-t
x=-45+t

left soliton
right soliton

b) Trajectories of the centers of QPs

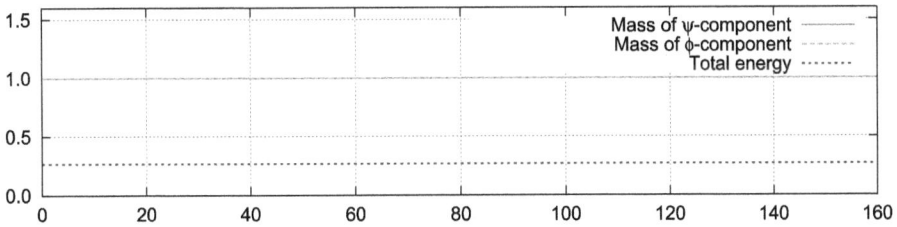

Mass of ψ-component
Mass of ϕ-component
Total energy

c) Masses and total energy as functions of time.

Figure 2.33. Head-on collision for $\Gamma = 0.1$. Linear initial polarization: $c_l = -c_r = 1$, $[\![\delta]\!] = 90°$.

beginning and 10^{-5} after the collision, i.e. it reasonably well approximates the trivial value. The small change is due to the reflection from the boundary conditions at the finite boundaries of the computational interval. The energy and the total mass are conserved, $E = 0.2659$, $M = 2$ at each time moment but negligible oscillations of the energy after the fifth significant digit and of the component masses M_ψ and M_ϕ after the seventh significant digit are present. The parameter that undergoes the more significant evolution is the polarization angle. We begin with linear

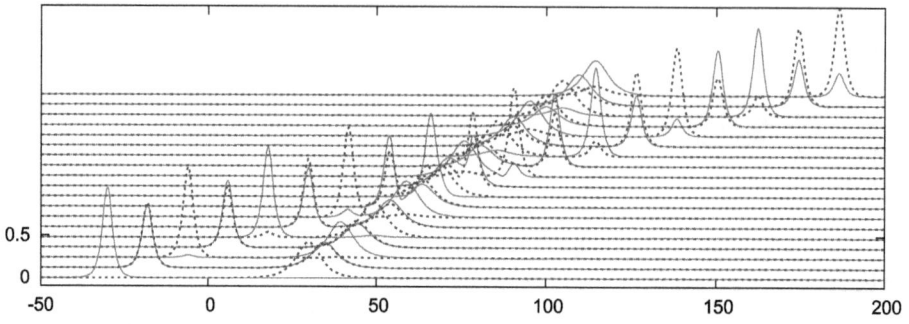

a) Profiles of the components

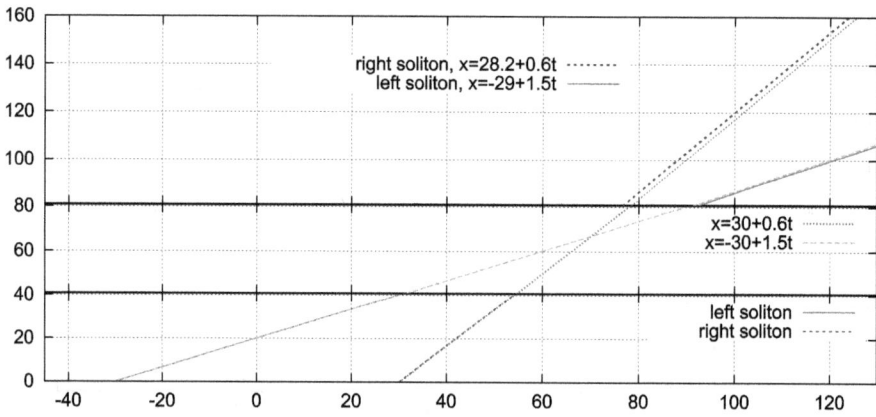

right soliton, x=28.2+0.6t ·······
left soliton, x=-29+1.5t ———

x=30+0.6t ·······
x=-30+1.5t ·······

left soliton ———
right soliton ·······

b) Trajectories of the centers of QPs

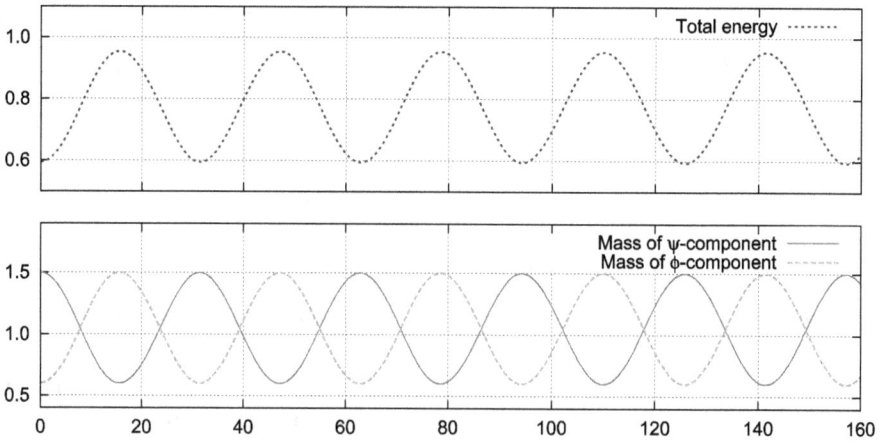

Total energy ·······

Mass of ψ-component ———
Mass of ϕ-component ·······

c) Masses and total energy as functions of time.

Figure 2.34. Takeover interaction for $\Gamma = 0.1$. Linear initial polarization: $c_l = 1.5$, $c_r = 0.6$, $[\![\delta]\!] = 0°$.

polarization ($\theta_1 = 0°$, $\theta_r = 90°$), which evolves after the interaction (see table 2.6, columns 2 and 3), but the sum is fairly well preserved (see table 2.6, column 4). The only place where the deviation of net polarization differs appreciably from 90° is at the moment of the interaction (see table 2.6, row 3). The approximate conservation of the polarization has also been found in our previous works with essential nonlinear coupling [100, 101].

The takeover interaction (figure 2.34(a)) follows the same qualitative pattern but with a bigger phase shift of the slower soliton and smaller phase shift of the faster soliton (figure 2.34(b)).

We note that the initial trajectories of centers shift from $x = -30 + 1.5t$ to $x = -29 + 1.5t$ and from $x = 30 + 0.6t$ to $x = 28.2 + 0.6t$ after the interaction. Compared to the previous case of the head-on collision, the interaction is longer and this is the reason for the larger phase shifts (see, for example, [79] with a similar effect for KdV solitons). We have found that the individual masses oscillate (breathe) by a period determined by the magnitude of the real part of the coupling parameter $\Re[\Gamma]$ (see figure 2.34(c)), while the net mass is constant at $M = 2.1$ (see table 2.7, column 7). Similarly, the total energy oscillates within the described period, but its magnitude is perfectly conserved within the full period (figure 2.34(c)). The pseudomomentum is also conserved at $P = 2.6056$ (see table 2.7, column 8). The notable feature here is that the rotational polarization originated by the linear coupling does not violate an excellent conservation of the total initial polarization $\theta_1 + \theta_r = 90°$ (see table 2.7, column 4).

Table 2.6. 'Breathing' dynamics of head-on interaction from figure 2.33.

t	θ_1	θ_r	$\theta_1 + \theta_r$	M_ψ	M_ϕ	$M_\psi + M_\phi$	P	E
0	0°	90°	90°	1	1	2	0.1×10^{-15}	0.2659
8	44° 9'	45° 50'	89° 59'	1	1	2	0.1×10^{-13}	0.2659
40	48° 33'	40° 15'	88° 48'	0.99	1	1.99	-0.5745×10^{-7}	0.2659
120	32° 28'	57° 31'	89° 59'	0.99	1	1.99	-0.2303×10^{-4}	0.2659
152	29° 08'	60° 52'	90°	0.99	1	1.99	-0.2294×10^{-4}	0.2659

Table 2.7. 'Breathing' dynamics of takeover interaction from figure 2.34.

t	θ_1	θ_r	$\theta_1 + \theta_r$	M_ψ	M_ϕ	$M_\psi + M_\phi$	P	E
0	0°	90°	90°	1.5	0.6	2.1	2.6056	0.5942378
8	45° 49'	44° 10'	89° 59'	1.037	1.063	2.1	2.6056	0.7794071
72	52° 27'	37° 29'	89° 56'	0.935	1.165	2.1	2.6056	0.8202960
120	57° 21'	32° 36'	89° 57'	1.238	0.862	2.1	2.6056	0.6989592
152	29° 18'	60° 43'	90° 1'	1.284	0.816	2.1	2.6056	0.6806916

The above discussion is concerned with the *sech*-like initial conditions when the linear-coupling parameter is real. In order to complete our investigation, we conduct a series of experiments with initial conditions of kinds (2.44a) and (2.44b) and complex-valued linear coupling. For simplicity, we use the same value for the real part, i.e. $\Gamma = 0.1 + 0.005i$. In figure 2.35(a), we present the case of head-on collision for linear initial polarization. The nontrivial imaginary part of the linear coupling leads to a violation of all observed conservation laws (figure 2.35(b)): the masses increase exponentially while keeping equal to one another; the (negative) total energy decreases exponentially too. The net pseudomomentum keeps its trivial value within good accuracy being in order of 10^{-17} in the beginning and 10^{-4} after the interaction (table 2.8, column 5).

Concerning the magnitudes of the net polarization of each envelope, we observed that they begin to breathe and gain amplitude right after the onset of time while keeping an excellent conservation of the total polarization before and after the interaction (see table 2.8, column 4).

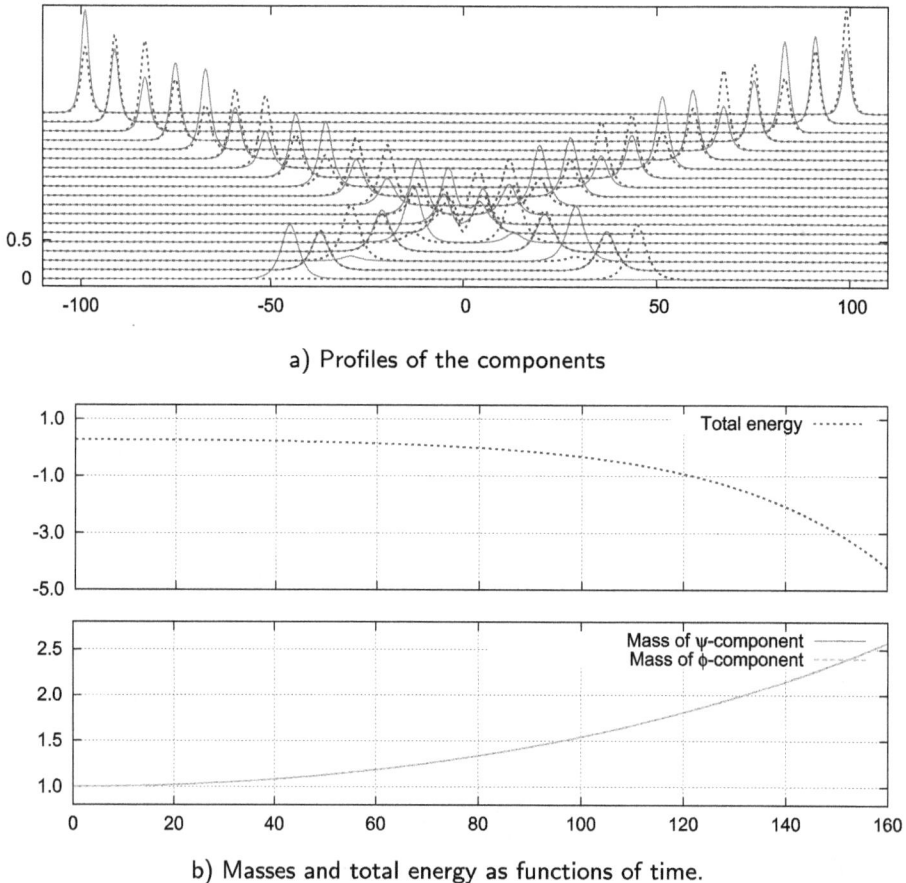

a) Profiles of the components

b) Masses and total energy as functions of time.

Figure 2.35. Head-on interaction for $\Gamma = 0.1 + 0.005i$. Linear initial polarization: $c_1 = -c_r = 1$, $[[\delta]] = 90°$.

Table 2.8. Complex (breathe and gain) dynamics of head-on interaction from figure 2.35.

t	θ_l	θ_r	$\theta_l + \theta_r$	P
0	$0°$	$90°$	$90°$	$0.73305489 \times 10^{-16}$
40	$40° \, 45'$	$48° \, 30'$	$89° \, 15'$	$-0.63825632 \times 10^{-4}$
80	$25° \, 35'$	$64° \, 25'$	$90°$	$-0.26786818 \times 10^{-4}$
120	$51° \, 41'$	$38° \, 20'$	$90° \, 01'$	$-0.17967968 \times 10^{-3}$
160	$55° \, 54'$	$34° \, 05'$	$89° \, 59'$	$-0.35057001 \times 10^{-3}$

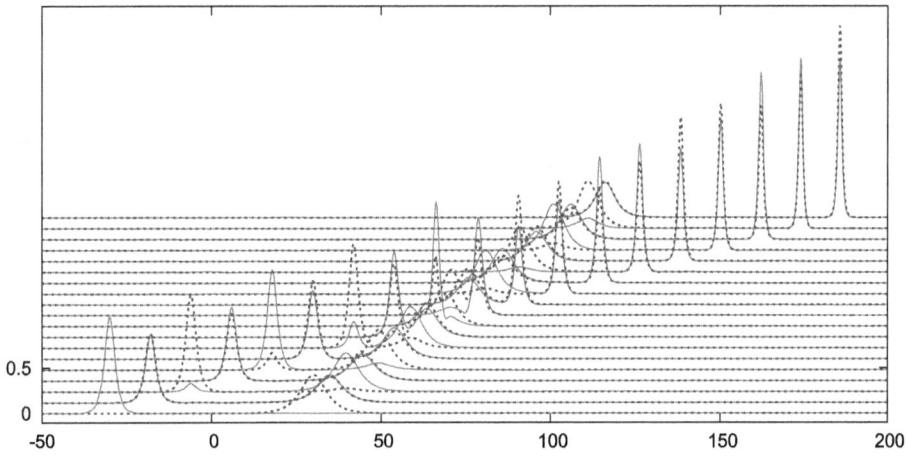

a) Profiles of the components

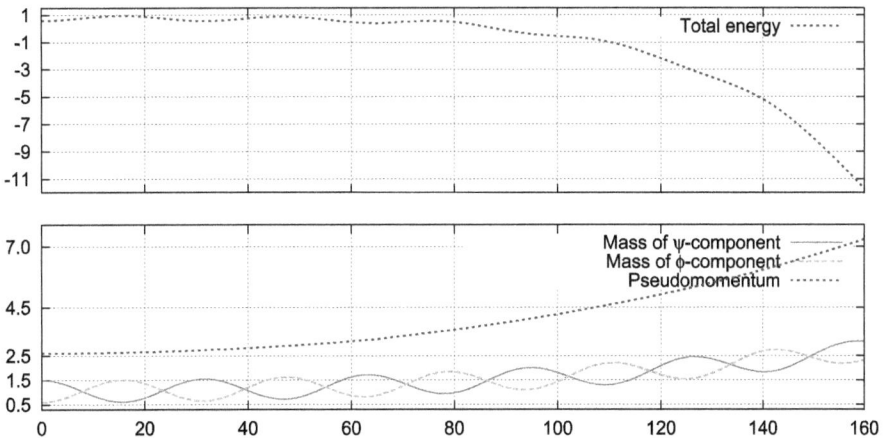

b) Masses, psedomomentum, and total energy as functions of time.

Figure 2.36. Takeover interaction for $\Gamma = 0.1 + 0.005i$. Linear initial polarization: $c_l = 1.5$, $c_r = 0.6$, $[[\delta]] = 90°$.

Table 2.9. Complex (breathe and gain) dynamics of takeover interaction from figure 2.36.

t	θ_1	θ_r	$\theta_1 + \theta_r$	P
0	0°	90°	90°	2.6056
16	85° 08′	4° 52′	90°	2.6390
40	41° 09′	48° 49′	90° 39′	2.8164
56	50° 04′	40° 48′	90° 52′	3.0234
80	52° 58′	54° 14′	107° 12′	3.5729
96	38° 37′	36° 45′	75° 22′	4.0749
120	48°	8° 43′	56° 43′	5.0271
136	42° 42′	77° 38′	120° 20′	5.8185
160	45° 16′	40° 13′	85° 29′	7.2992

In the end, we consider the case of takeover collision with the complex parameter of linear coupling. Compared to the linear coupling with purely real coupling parameter (figure 2.34), we again observe oscillations of the energy and masses (figure 2.36(b)). The (negative) energy decreases very fast, while the masses M_ψ and M_ϕ increase all of them oscillating. The pseudomomentum P increases without appreciable oscillation. Table 2.9 presents the polarization. It is seen that the individual polarizations oscillate from the very beginning and the net total polarization is conserved only prior to the interaction. After the interaction, the total polarization starts to oscillate too (see table 2.9, column 4). It is still an open question whether the polarization is conserved over the period of the oscillations.

Conclusion

In the present note, we solve numerically VNLSEs with complex-valued linear coupling. We find that in the case of a purely real linear coupling parameter, the energy, pseudomomentum, mass, and polarization are conserved over the period of oscillations (breathing), while adding an imaginary part to the parameter leads to a violation of the conservation laws. Our computations confirm the well established fact for many different soliton systems that the phase shift depends on the duration of the interaction being larger for takeover collisions in comparison with the head-on collision.

We study the polarization dynamics for solitons taking over one another, i.e. moving in one direction with different velocities [104]. The initial polarizations are elliptic. To 'extract' the pure effect of the initial phase difference, we consider in the beginning the MS ($\alpha_2 = 0$). The total masses, pseudomomenta, and energy of both OPs are kept constant. The combination of nontrivial nonlinear coupling (cross modulation α_2) and nontrivial initial phase difference influences in phase shift of the velocities of interacting solitons.

2.9 Polarization dynamics during takeover collisions of solitons in vector nonlinear Schrödinger equation

In the case of general elliptic polarization, the analytical solution for the shapes of steadily propagating solitons are not available, and we develop a numerical algorithm finding the shape. The results of this work outline the role of the initial phase, initial polarization and the interplay between them, as well as the nonlinear couplings on the interaction dynamics of solitons in VNLSE.

The VNLSE is a soliton supporting system appearing as a model for light propagation in isotropic Kerr materials [25, 54, 75, 87]. There are many different solitons supporting systems whose solutions behave as QPs, but NLSE and VNLSE remain two of the most important featuring models for the quasi-particle concept. For this reason, the VNLSE attracted the attention of leading researchers, and a number of excellent analytical results have been obtained over the years. For the integrable Manakov case (see [5, 18, 26, 60, 72, 74, 86], among others), linear and circular polarization were treated in [69], and general polarization in [55, 112], among others.

Numerical method

The time-stepping scheme needs initial conditions of the type of equation (2.24). In the cases of linear and circular polarization, the functions A_χ in equation (2.24) are *sech*-es. In the general case of elliptic initial polarization, the carrier frequencies $n_\psi \neq n_\phi$, and we discretize the system equations (2.37) in the same manner as the evolutionary system, and use Newton's method and Hermitian splines to obtain the solution. The details can be found in [102]. Our results are confirmed [95] and can be used with confidence as initial conditions for the problem under consideration.

Takeover collisions: results and discussion

Manakov solutions
The takeover collision of solitons is another basic interaction alongside with the head-on collision. The difference between the two kinds of collision is that the interaction between quasi-particles (QPs) can be longer during the takeover collision. Our goal is to investigate such kinds of interaction and to understand the influence of the initial polarization and initial phase difference on the behavior of the outgoing QPs after the interaction. We start with linear initial polarization when we have exact (*sech*) envelopes. Let us consider Manakov solutions when the nonlinear coupling vanishes and elastic interactions are present. We conduct our investigation for given phase speeds of the envelopes, say $c_l = 1.5$ and $c_r = 0.6$, which are the phase speeds of the left and right incoming soliton, respectively. The interaction is fully elastic and the phase speeds of the outgoing solitons are the same as the incoming speeds. Moreover, we find out that the initial phase difference $[\![\delta]\!] \equiv \delta_r - \delta_l$ does not affect the elasticity of interaction (figure 2.37). In all these observations, the masses, pseudomomentum and energy are conserved before and after the interaction. Keeping in mind the quadratic integrands for the masses and

the energy of QPs in the head-on collision are the same and only the pseudomomentum varies (see table 2.10, columns e–h). In figure 2.37(b), we note that the centers of QPs keep their velocities but are shifted slightly after the interaction: the faster soliton (to the right) by 1.0 units from $x = -30 + 1.5t$ to $x = -29 + 1.5t$ while

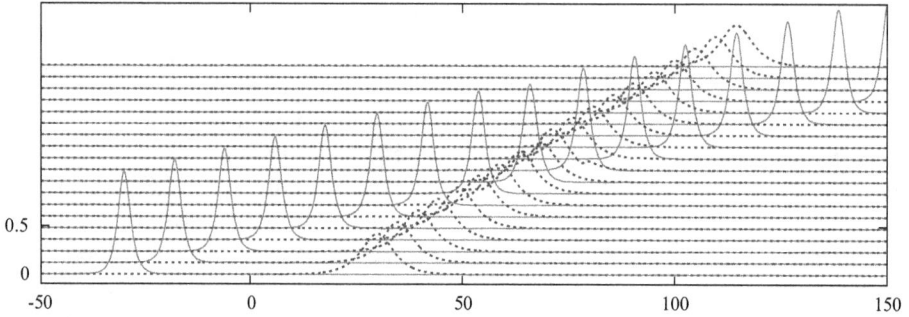

(a) Profiles of interaction

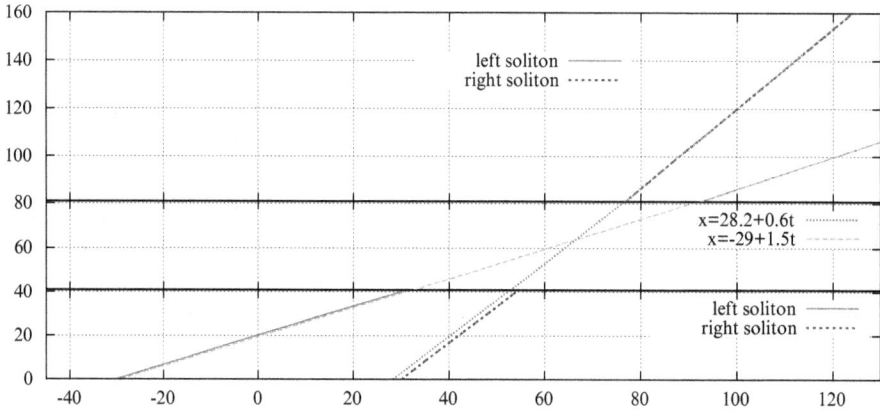

(b) Trajectories of soliton centers

Figure 2.37. Takeover collision. Elastic interaction with initial linear polarization, $\alpha_2 = 0$, $c_l = 1.5$, $c_r = 0.6$, $[[\delta]] = 90°$.

Table 2.10. Polarization evolution during takeover collisions. Initial linear polarization.

				(a) Initial values.				
$[[\delta]]$	θ_l^i	θ_r^i	$\theta_l^i + \theta_r^i$	m_ψ^i	m_θ^i	p^i	e^i	α_2
90°	0°	90°	90°	1.5	0.6	2.60565	1.1942321	0
0°	0°	90°	90°	1.5	0.6	1.88589	1.1942321	0

				(b) Values after the interaction.				
$[[\delta]]$	θ_l^f	θ_r^f	$\theta_r^f + \theta_r^f$	m_ψ^f	m_ϕ^f	p^f	e^f	α_2
90°	0°	90°	90°	1.5	0.6	2.60565	1.1942321	0
0°	0°	90°	90°	1.5	0.6	1.88589	1.1942321	0

the slower soliton (to the left) shifts by 1.8 units from $x = 30 + 0.6t$ to $x = 28.2 + 0.6t$. As one can see in figures 2.38(a) and (b), these results coincide qualitatively with our previous results for head-on collisions where the center of the left incoming soliton moves from line $x = -45 + 1.5t$ to line $x = -44.5 + 1.5t$ and the right incoming soliton—from $x = 45 - 0.6t$ to $x = 44 - 0.6t$.

In order to complete the results with initial *sech*-like envelopes, we consider the same problem but with circular and elliptic initial conditions. In these cases, the shape of the initial profile has a more general form and there exists no known exact solutions to be used for this purpose. In [100, 102], we derived initial envelopes as the difference solution of a coupled nonlinear system of ODEs. We use them as approximate initial conditions in the case of circularly and elliptically polarized initial conditions. For simplicity, we fix the initial velocities of envelopes to be $c_l = 1$ and $c_r = 0.8$. If the polarization is circular, the carrier frequencies are equal, $n_\psi = n_\phi = -1.5$ (figure 2.39(a) and (b)) while if the polarization is elliptic—$n_\psi = -1.1$, $n_\phi = -1.5$ (figure 2.40(a)). In all of the computed cases, we observe a phase shift of

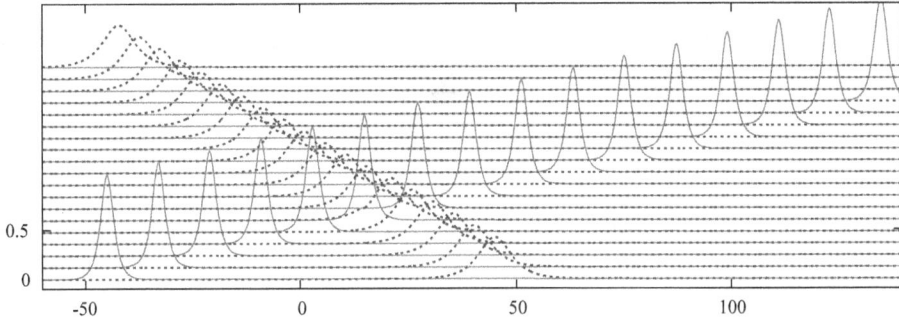

(a) Profiles of interaction

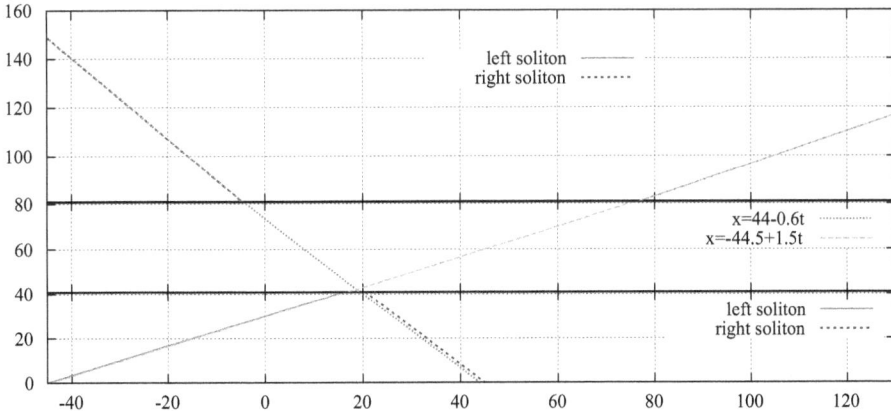

(b) Trajectories of soliton centers

Figure 2.38. Head-on collision. Elastic interaction with initial linear polarization, $\alpha_2 = 0$, $c_l = 1.5$, $c_r = -0.6$, $[[\delta]] = 90°$.

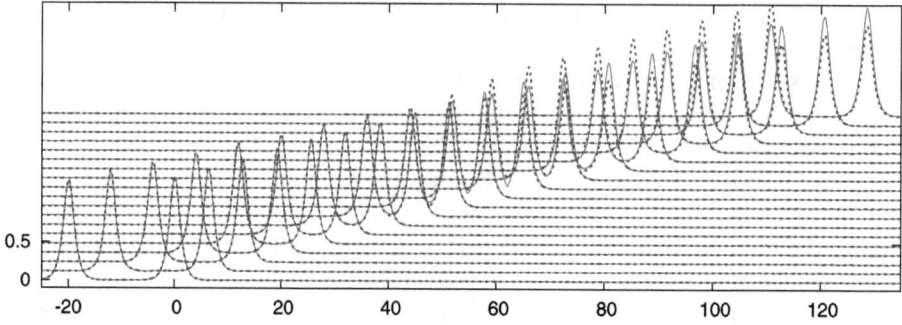

(a) Profiles of interaction for $[\![\delta]\!] = 0°$

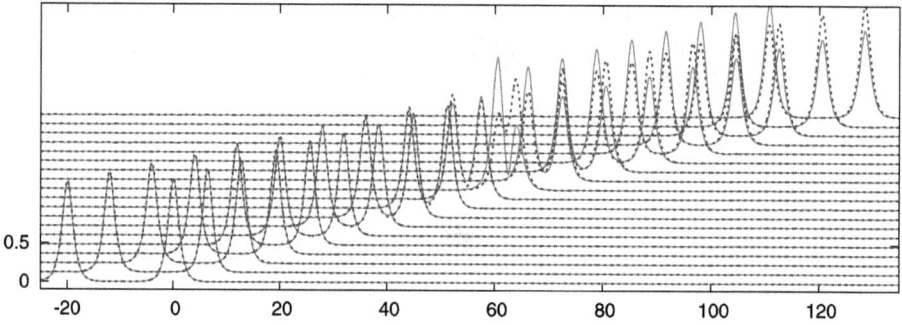

(b) Profiles of interaction for $[\![\delta]\!] = 45°$

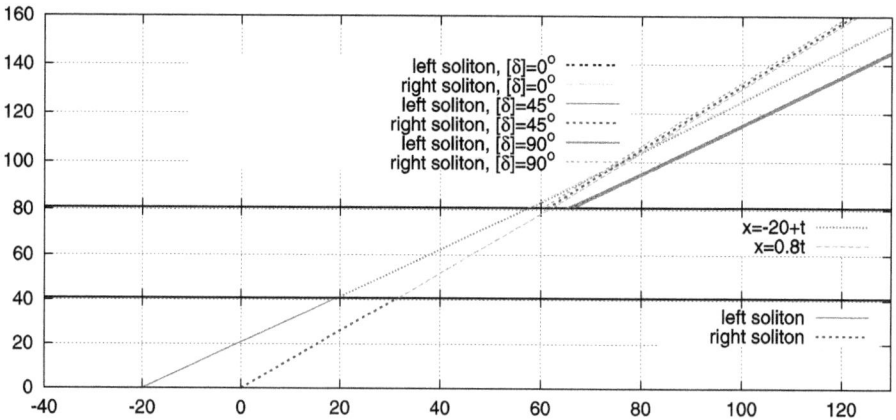

(c) Trajectories of soliton centers for $[\![\delta]\!] = 0°; 45°; 90°$

Figure 2.39. Takeover collision. Elastic interaction with initial circular polarization, $\alpha_2 = 0$, $c_l = 1$, $c_r = 0.8$, $n_\psi = n_\phi = -1.5$.

the outgoing solitons combined with no career velocity change (figures 2.39(c) and 2.40(b)). Obviously, the nontrivial initial phase difference $[\![\delta]\!]$ can only contribute to the phase shifts making them larger or smaller but always the faster QP (with smaller mass) suffers a larger shift compared to the slower QP (with larger mass), which can

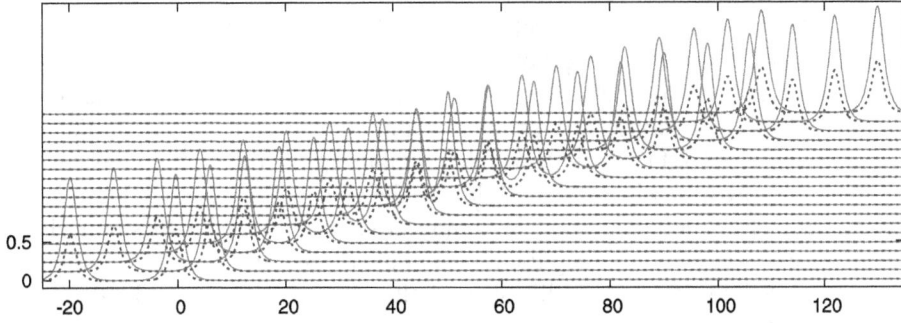

(a) Profiles of interaction for $[\![\delta]\!] = 0°$

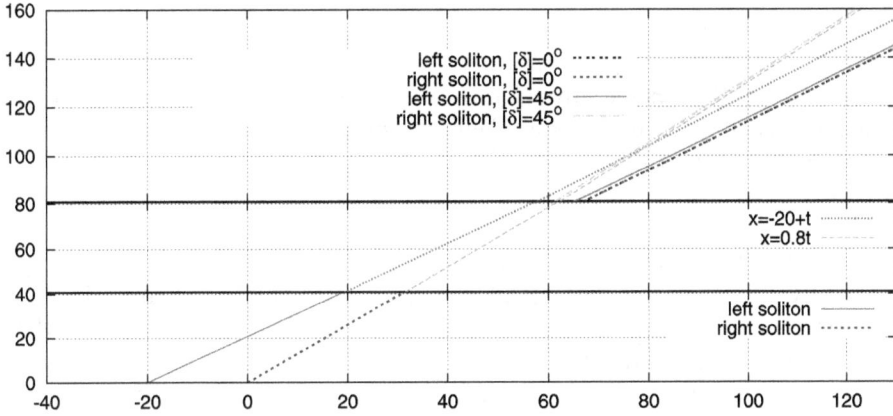

(b) Trajectories of soliton centers for $[\![\delta]\!] = 0°; 45°$

Figure 2.40. Takeover collision. Elastic interaction with initial elliptic polarization, $\alpha_2 = 0$, $c_1 = 1$, $c_r = 0.8$, $n_\psi = -1.1$, $n_\phi = -1.5$.

Table 2.11. Polarization evolution during takeover collisions. General initial polarization.

(a) Initial values.

$[\![\delta]\!]$	θ_l^i	θ_r^i	$\theta_l^i + \theta_l^i$	m_ψ^f	m_ϕ^f	p^i	e^i	α_2
0°	45°	45°	90°	3.033	3.033	5.44	−2.7633	0
45°	45°	45°	90°	3.033	3.033	5.44	−2.7633	0
0°	26° 37′	24° 29′	51° 6′	3.816173	0.735471	4.08697	−1.05586	0
45°	26° 37′	24° 29′	51° 6′	3.816173	0.735471	4.08697	−1.05586	0

(b) Values after the interaction.

$[\![\delta]\!]$	θ_l^f	θ_r^f	$\theta_l^f + \theta_r^f$	m_ψ^f	m_ϕ^f	p^f	e^f	α_2
0°	50° 53′	39° 47′	94° 26′	3.033	3.033	5.44	−2.7633	0
45°	52° 7′	38° 56′	91° 3′	3.033	3.033	5.44	−2.7633	0
0°	23° 39′	25° 57′	49° 36′	3.816173	0.735471	4.08697	−1.05586	0
45°	26° 55′	20° 7′	47° 2′	3.816173	0.735471	4.08697	−1.05586	0

suffer two to three times smaller shift (see table 2.11, columns e and f). In all of the considered cases, we have excellent conservation of the masses of ψ- and ϕ-components, as well as the full pseudomomentum and energy (see table 2.11, columns e–h). These conserved properties do not depend on the initial phase shift $[\![\delta]\!]$ (see table 2.11, column a). We establish that the total polarization of envelopes is approximately conserved during the elastic interaction. The individual polarizations of the left and right incoming envelopes give rise to slight changes during the takeover but they keep their initial kinds: circularly polarized QP remains circularly polarized after the collision and the elliptically polarized one remains elliptically polarized. This observation confirms one more time that the Manakov solution is not unique (see [101]).

Nonlinear coupling

Allowing a nonlinear coupling (nontrivial cross-modulation parameter α_2) transforms the interaction, essentially. In this case, the system loses its full integrability. We will consider the general case of elliptic initial polarization with polarization angles $\theta_l = 39° \, 51'$ and $\theta_r = 40° \, 17'$. We choose $\alpha_2 = 2$.

Conclusions

The polarization dynamics of takeover vector solitons of the VNLSE is investigated numerically by means of an energy conserving difference scheme. The initial condition is obtained numerically by solving the conjugate system of ODEs for the shape of the stationary propagating vector soliton. For the latter, the carrier frequencies of the two components of the vector soliton are different and the amplitude of the envelope is no longer a *sech*. Different combinations of elliptically polarized (e.g. circularly polarized for the particular case of angle of polarization $\theta = 45°$, linearly polarized $\theta = 0°$; $90°$) solitons are used as the initial condition and the evolution of the system is followed numerically. Using a general initial polarization significantly enriches the observed effects of the nonlinear coupling in VNLSE (2.41). Our results show that the polarization dynamics is very susceptible to the initial phases of the solitons.

First, we investigate the case of trivial cross-modulation, $\alpha_2 = 0$. Our computations show that the initial phase difference $[\![\delta]\!]$ generates a phase shift of the soliton centers but a change of phase speed shift is not observed as a rule. The interaction is fully elastic and qualitatively is quite similar to the head-on collision. However, the takeover collision is more prolonged compared to the head-on one and this is the reason for larger phase shifts to accumulate. Similar arguments hold for the takeover collision dynamics of the solutions of the KdV equation as well [7]. Solutions (generally not *sech*-like) with a polarization change persist after the interaction as QPs. The latter means that a bifurcation takes place during the cross-section of the interaction keeping the functions smooth and breaking the individual polarization magnitudes (a kind of 'shock') but conserving the total magnitude of polarization of the QPs configuration.

Second, we treat cases with nontrivial cross-modulation, $\alpha_2 \neq 0$. We find that the actual magnitude of α_2 together with the initial phase difference $[\![\delta]\!]$ both have a

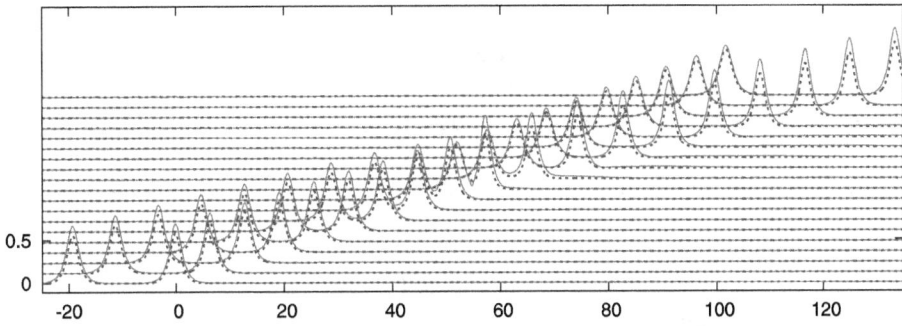

(a) Profiles of interaction for $[\![\delta]\!] = 0°$

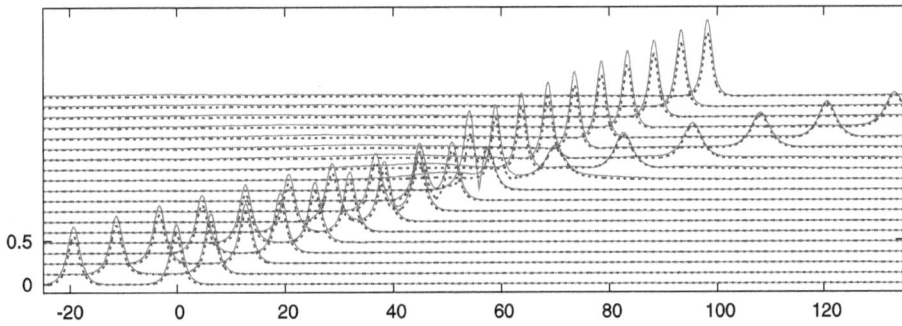

(b) Profiles of interaction for $[\![\delta]\!] = 45°$

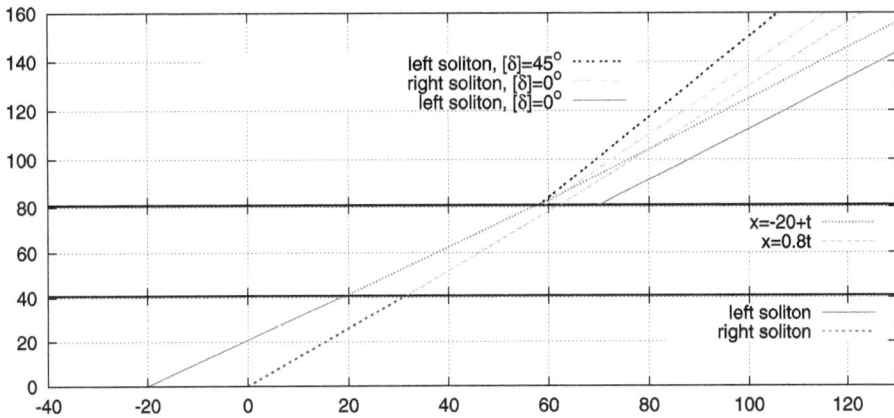

(c) Trajectories of soliton centers for $[\![\delta]\!] = 0°; 45°$

Figure 2.41. Nonlinear takeover collision, $\alpha_2 = 2$. An interaction with initial elliptic polarization, $c_1 = 1$, $c_r = 0.8$, $n_\psi = -1.1$, $n_\phi = -1.5$.

qualitative and a quantitative effect on the behavior of the interacting and outgoing solitons. Our results show that the combined effect of these two parameters inevitably transform the initial polarization regardless of it being linear, circular or elliptic to elliptic polarization of the outgoing solitons, but always conserving its total value of the polarization. The phase shift is accompanied by a change of phase speed, depending on the initial phase difference $[[\delta]]$. Thus, one of the main results of the present paper is in identifying the role of the initial phase in the case of takeover collisions, showing that it can have a substantial effect on the interaction for non-trivial cross-modulations.

VNLSE, in opposition to the scalar NLSE, is not fully integrable and admits two conservation laws—for the mass and pseudomomentum, as well as a law for the balance of the energy. For flat asymptotic conditions, the last one becomes one more conservation law. In all of the cases considered by us, the discrete analogues of these laws hold approximately within the round-off error due to the conservative difference scheme used. The same conclusion relates to the total polarization of the QP configuration—the sum of individual polarization angles is a constant. Though, in the place of the interaction, the curves of individual polarization have jumps, their sum remains unchanged. These results argue us to assume that one more net conservation law holds—for the total polarization. The involvement of the linear coupling gives rise to rotational polarization and (breathing solitons) breathers regardless of the independence of the initial polarization and phase difference [103, 106]. Yet, our investigations demonstrate [106] that in this complicated case the net conservation law for the total polarization is liable for a generalization—in contrast to the individual and total masses (see [103]), the individual polarization angles phase shift one another and this is the reason both the individual polarizations and the total polarization breathe [106].

2.10 The effect of the elliptic polarization on the quasi-particle dynamics of linearly coupled vector nonlinear Schrödinger equation

We investigate numerically by a conservative difference scheme in complex arithmetic the head-on and takeover collision dynamics of the solitary waves as solutions of linearly coupled NLSEs for various initial phases. The QP behavior of propagating and interacting solutions in conditions of elliptic and rotational polarizations at the same time is examined.

The VNLSE is attractive for several reasons. Among them are: VNLSE is a soliton supporting system; the system is a proper 'spring board' for investigating and tracking the QP behavior of the solitary waves; it possesses very rich phenomenology which can be extended by adding new terms in the equations; and the vector structure of the system and its solutions permit us to study and treat the polarization vector [55, 69, 112]. The initial polarization of VNLSE and its evolution during the QP interaction is a very important element. There are numerous experimental observations (see [24] and literature cited therein) of the effects connected with this quantity. For the fully fledged VNLSE with general initial polarization, no analytic solution is available. It can be obtained, e.g. via an adequately devised numerical scheme. Keeping in mind the importance of initial polarization, we investigated in [101] the collision dynamics for circularly polarized solitons based on *sech*-functions and found out that there exists an infinite number of Manakov two-soliton solutions preceded by polarization discontinuity (shock) on the place of interaction. In order to extend the range of investigations in the case of general polarization, we established an auxiliary conjugate system of nonlinear ordinary differential equations, equations (2.37), in order to generate numerically initial elliptically polarized soliton solutions and applied them for numerous head-on [105] and takeover collisions [104]. To enrich the concept of the polarization, we took into consideration the linear coupling of *sech*-like solitons, which is the generator of their rotational polarization (breathing) and/or gain/self-dissipation [103].

Here, we conduct a series of simulations in order to track the particle-like behavior of the interacted localized non-*sech* waves; the latter having arbitrary (elliptic) and rotational initial polarizations at the same time. By studying the collision dynamics of the QPs, we aim to investigate the interaction of both kinds of polarizations in the general case of non-*sech*-like solutions. Let us clarify that the takeover investigation conducted above concerned only linearly polarized breathers that set limits on the consideration of *sech*-like solutions only. Our final goal is to generalize the net law of the polarization dynamics. For completeness, we pay attention to the arbitrary (elliptic) and the rotational polarizations at the same time, as well as to the circular (as a special kind of elliptic) and rotational polarizations at the same time.

Problem formulation

VNLSE (2.21a) and (2.21b) is a system of nonlinear wave equations (also called the Gross–Pitaevskii or Manakov-type system). When $\alpha_2 = 0$, no nonlinear coupling is present, despite the fact that 'cross-terms' proportional to α_1 appear in the equations and both of them admit an identical solution equal to the solution of a single NLSE with the nonlinearity coefficient $\alpha = 2\alpha_1$. Let us pay attention to the modified VNLSE where the nonlinear coupling is trivial, but is replaced by the so-called linear coupling. Nevertheless, the system, equations (2.21a) and (2.21b), remains nonlinear and gets the form

$$i\psi_t = \beta\psi_{xx} + \alpha_1(|\psi|^2 + |\phi|^2)\psi - \Gamma\phi, \tag{2.47a}$$

$$i\phi_t = \beta\phi_{xx} + \alpha_1(|\phi|^2 + |\psi|^2)\phi - \Gamma\psi. \tag{2.47b}$$

The magnitude of linear coupling is governed by the complex-valued quantity Γ. $\Re\Gamma$ governs the oscillations between states termed as breathing solitons and actually it proves to be their frequency of oscillation, while $\Im\Gamma$ describes the gain/self-dissipation behavior of soliton solutions. As is shown in [92, 103], the breathing is presently free of the interaction. The relation between the Manakov-type system, equations (2.21a) and (2.21b), (when $\alpha_2 = 0$) and the linearly CNLSE, equations (2.47a) and (2.47b), is given by the substitution (for details, see [92] and [103])

$$\psi = \Psi\cos(\Gamma t) + i\Phi\sin(\Gamma t), \quad \phi = \Phi\cos(\Gamma t) + i\Psi\sin(\Gamma t), \tag{2.48}$$

where Φ and Ψ are solutions of equations (2.21a) and (2.21b). Let us emphasize that the modified system, equations (2.47a) and (2.47b), is not a Manakov-type system and due to the linear coupling term possesses more rich phenomenology. Both systems are integrable and complete integrability is reached only for $\alpha_2 = \Gamma = 0$. In this case, the systems, equations (2.21a) and (2.21b) and equations (2.47a) and (2.47b), are identical and coincide with the scalar NLSE (for more details, see, for example, [92]). These properties and references (2.48) of the solutions allow us to solve equations (2.47a) and (2.47b) instead of equations (2.21a) and (2.21b) in order to get more information about the breathing behavior of the soliton solutions. Let us remind you that in [103], we consider the problem in question but in the particular case when Φ and Ψ were assumed to be certainly $sech$-solutions of (2.21a) and (2.21b).

The solutions under consideration are pulses whose modulation amplitude is of a general form (non-$sech$) and their polarization rotates with time. This determines the choice of the initial conditions for numerical investigation of temporal evolution of interacting solitons. As usual, we are concerned with the soliton solutions which are localized envelopes on a propagating carrier wave. We assume that for each of the functions ϕ, ψ, the initial condition has the general type, equation (2.46), with the usual notations. We are interested in solutions with initially nontrivial carrier frequencies $(n_\psi, n_\phi) \neq (0, 0)$, i.e. with elliptic or circular polarizations. Let us emphasize that these qualities are intrinsic to non-$sech$-like solutions. Keeping in mind (2.46), we implement straightforward manipulations, yielding

$$\left[A_\psi'' + \left(n_\psi + \frac{1}{4}c^2 \right) A_\psi + \alpha_1 \left(A_\psi^2 + A_\phi^2 \right) A_\psi \right] \exp \left\{ i \left[n_\psi t - \frac{1}{2}c(x - X - ct) + \delta_\psi \right] \right\} \cos(\Gamma t)$$

$$+ i \left[A_\phi'' + \left(n_\phi + \frac{1}{4}c^2 \right) A_\phi \right. \tag{2.49a}$$

$$\left. + \alpha_1 \left(A_\phi^2 + A_\psi^2 \right) A_\phi \right] \exp \left\{ i \left[n_\phi t - \frac{1}{2}c(x - X - ct) + \delta_\phi \right] \right\} \sin(\Gamma t) = 0,$$

$$\left[A_\phi'' + \left(n_\phi + \frac{1}{4}c^2 \right) A_\phi + \alpha_1 \left(A_\phi^2 + A_\psi^2 \right) A_\phi \right] \exp \left\{ i \left[n_\phi t - \frac{1}{2}c(x - X - ct) + \delta_\phi \right] \right\} \cos(\Gamma t)$$

$$+ i \left[A_\psi'' + \left(n_\psi + \frac{1}{4}c^2 \right) A_\psi \right. \tag{2.49b}$$

$$\left. + \alpha_1 \left(A_\psi^2 + A_\phi^2 \right) A_\psi \right] \exp \left\{ i \left[n_\psi t - \frac{1}{2}c(x - X - ct) + \delta_\psi \right] \right\} \sin(\Gamma t) = 0.$$

The above two equations are true if and only if the two components of the envelope are governed by the following coupled system of differential equations:

$$A_\psi'' + \left(n_\psi + \frac{1}{4}c^2 \right) A_\psi + \alpha_1 \left(A_\psi^2 + A_\phi^2 \right) A_\psi = 0, \tag{2.50a}$$

$$A_\phi'' + \left(n_\phi + \frac{1}{4}c^2 \right) A_\phi + \alpha_1 \left(A_\phi^2 + A_\psi^2 \right) A_\phi = 0. \tag{2.50b}$$

We find out that this system is explicitly equivalent to the bifurcation conjugate system corresponding to the Manakov case with $\alpha_2 = 0$ (see [105] and [102]). Here, we aim to trace the strong linear coupling on the interaction in the conditions of general initial polarization. After that, we construct the initial conditions for $\vec{\chi}$ from (2.46) as a superposition of two two-component solitary waves located at positions X_l and X_r and propagating with phase speeds c_l and c_r. The initial distance $|X_l - X_r|$ is supposed to be large enough in order for the solutions not to overlay in the time moment $t = 0$. The system, (2.50a) and (2.50b), is solved numerically and their solutions are readily used as an approximate initial condition for the unsteady computations. Our aim is to better understand the influence of the initial polarization and the initial phase difference on the particle-like behavior of the localized waves (QPs). They survive the collision with other QPs (or some other kind of interactions) without losing its identity. The initial difference in phases can have a profound influence on the polarizations of the solitons after the interaction, as well as on the magnitude of the full energy and the amplitude of breathing.

Results and discussion

We conducted numerous numerical simulations tracking the two main interactions—head-on and takeover collisions of stationary and breathing solitons. The main goal was to understand the influence of the initial polarization on the mechanism of interactions. For the big parametric set and rich phenomenology of the system of CNLSE (VNLSE), we resume the essential results and observations.

Circular polarization: head-on collision
Circular polarization is a special elliptic polarization with polarization angle $\theta = 45°$. The initial configuration is generated from the auxiliary bifurcation system, (2.50a) and (2.50b). Since the parametric space of the problem is too big to be explored in full, we choose $n_{l\psi} = n_{r\psi} = n_{l\phi} = n_{r\phi} = -1.5$, $c_l = -c_r = 1$, $\alpha_1 = 0.75$, $\Gamma = 0.175$ and focus on the effects of $[[\delta]] \equiv \delta_r - \delta_l$. We observe that the breathing does not interfere with the soliton collision since they are distinguished from previously obtained breathers [20, 111]. One sees the 'breathing' of the pulses even without any interaction. This is the manifestation of the rotation of the polarization. Thus, Γ is responsible for the exchange of wave mass between the modes and we call it 'cross' dispersion of the signals (see figure 2.43). Recall that these solutions apply specifically to the case $\alpha_2 = 0$. Our initial condition is valid only for this case. In figure 2.42, we present these features for a fixed set of parameters varying only the initial phase difference $[[\delta]]$. As we discussed in [103], there is no velocity shift on the place of interaction. The individual masses start to oscillate since the same beginning with an equal period and with a phase shift and an amplitude depending on the concrete choice of $[[\delta]]$. The total mass, however, is perfectly conserved—it is equal to 6.0683. Due to the symmetry, the magnitude of the net pseudomomentum P varies in the interval $(10^{-11}, 10^{-3})$ and evidently approximates 0. The result is the same for all considered phase differences (0°, 90°, 180°) (figure 2.42). The magnitude of the energy E essentially depends on the magnitude of the initial phase difference

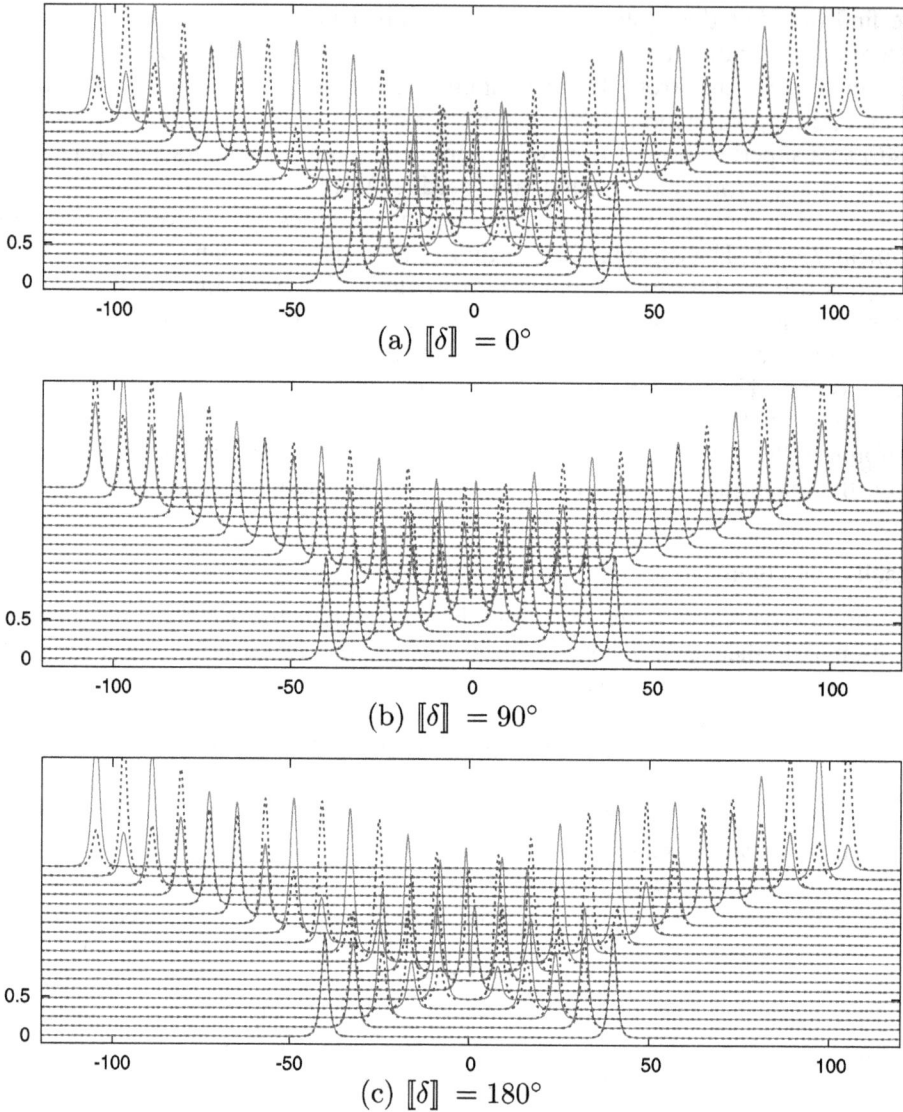

Figure 2.42. Head-on collision of QPs (ψ–solid line; ϕ–dashed line) with initial circular polarization for $n_{l\,\psi} = n_{r\,\psi} = n_{l\,\phi} = n_{r\,\phi} = -1.5$, $c_l = -c_r = 1$, $\alpha_1 = 0.75$, $\Gamma = 0.175$, $X_l = -40$, $X_r = 40$ and different initial phase differences.

value $[[\delta]]$. The dependence is non-monotonous. For example, $E = -3.078$ when $[[\delta]] = 0°$; $E = -4.15$ when $[[\delta]] = 90°$; $E = -1.139$ when $[[\delta]] = 180°$. The next figure, 2.44, elucidates the polarization evolution. The individual polarizations oscillate depending on both the magnitude of linear coupling Γ and the initial phase difference $[[\delta]]$. These oscillations do not depend on the interaction but the period depends on Γ. The magnitude $[[\delta]]$ influences the amplitude of oscillation. In contrast to masses (see figure 2.43), the total polarization is not constant. It also breathes,

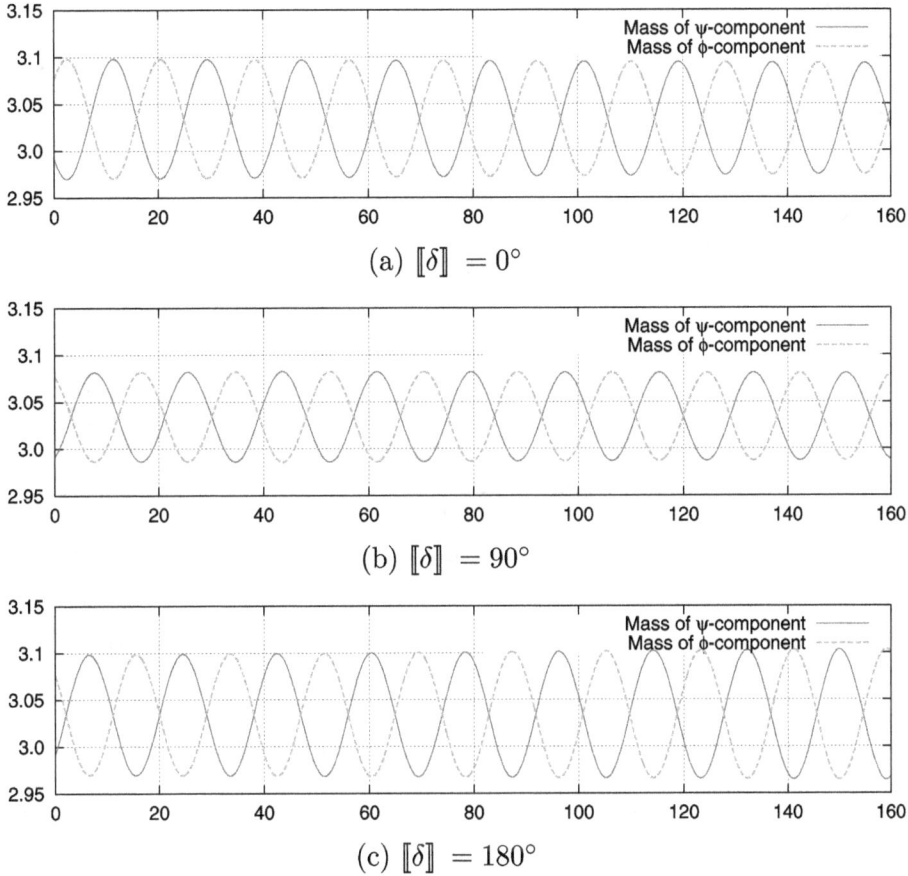

(a) $[\![\delta]\!] = 0°$

(b) $[\![\delta]\!] = 90°$

(c) $[\![\delta]\!] = 180°$

Figure 2.43. Mass dynamics, corresponding to figure 2.42.

though with very small amplitude. It is well visible that in the place of an interaction, one observes a shock of both the individual and the total polarization. The last result can be interpreted as a generalization of the polarization conservation law for the case of rotational polarization—it breathes in the period but it is conserved in the whole period.

Our observations show that these interactions are independent of initial phase shift and no-phase velocity shift occurs on the place of interaction. The trajectories of QP-centers suffer a slight shift. They were investigated and discussed in [103] for *sech*-profiles and initial linear polarization, and we do not concern ourselves with them again.

Elliptically polarized solitons: head-on collision
In order to extend our considerations, we conduct a series of experiments with general initial polarization and nontrivial linear coupling. For convenience, we use the same magnitude of $\Gamma = 0.175$. In figure 2.45, the case of initial elliptic polarization is present for initial phase shifts $[\![\delta]\!] = 0°$, $90°$, $135°$, $180°$.

(a) $[\![\delta]\!] = 0°$

(b) $[\![\delta]\!] = 90°$

(c) $[\![\delta]\!] = 180°$

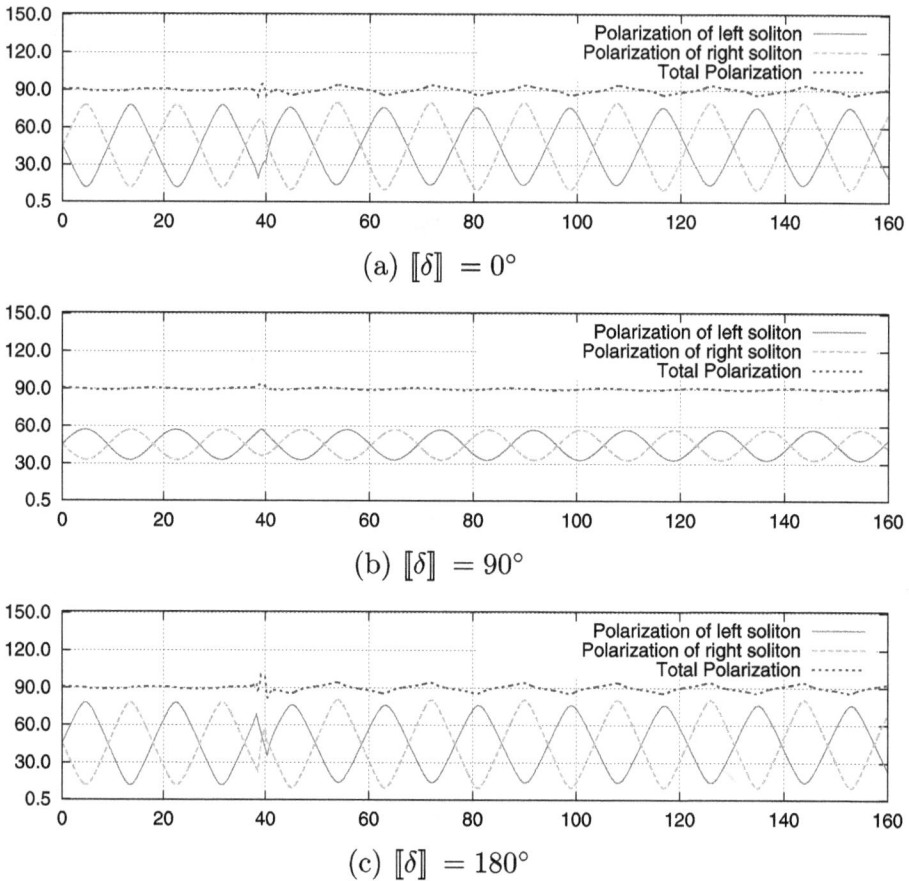

Figure 2.44. Individual and total polarization dynamics in dependence on initial phase difference, corresponding to figure 2.42.

Our conclusion is that the linear coupling combined with the initial phase difference $[\![\delta]\!]$ can only shift the soliton trajectories and keeps the phase velocities unchanged. A nontrivial phase shift is possible after involving nonlinear coupling $\alpha_2 \neq 0$ [104]. We consider two solitons with equal initial elliptic polarization angles of $23° \, 44'$. The initial configuration is generated from the auxiliary bifurcation system, (2.50a) and (2.50b). The above angle corresponds to the parametric set $n_{l\psi} = n_{r\psi} = -1.1$, $n_{l\phi} = n_{r\phi} = -1.5$, $c_l = -c_r = 1$, $\alpha_1 = 0.75$. Obviously, the results are qualitatively the same as in the previous case of circular polarization and they do not require a detailed discussion. The correlation between the energy and the initial phase difference is essential: for $[\![\delta]\!] = 0° - E = -0.262$; for $[\![\delta]\!] = 90° - E = -0.821$; for $[\![\delta]\!] = 135° - E = -0.206$; for $[\![\delta]\!] = 180° - E = 0.640$. The breathing behavior of the total polarization is well visible, as well as the effect of the initial phase difference upon the amplitude of the individual polarization angles, which affect the magnitude of the total polarization angle. Figure 2.45 convincingly answers the question about the dynamics of the rotational polarization when the type of initial

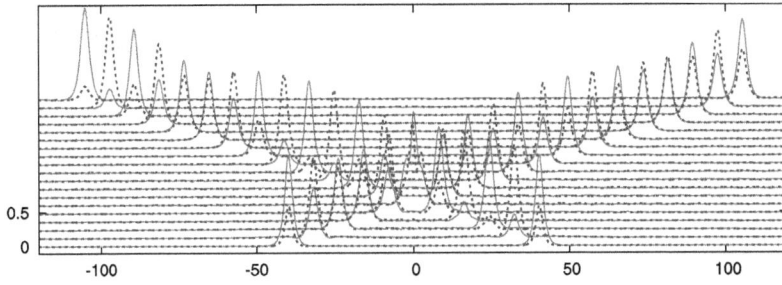

(a) $[\![\delta]\!] = 0°$

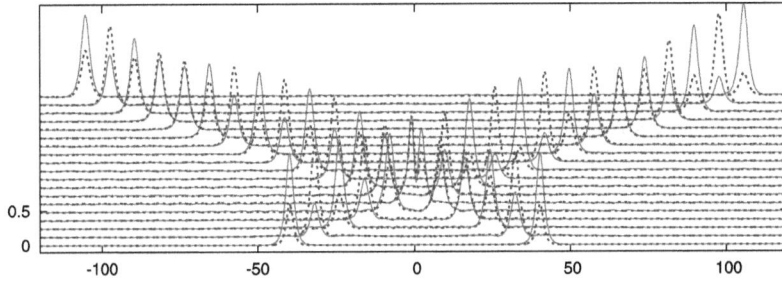

(b) $[\![\delta]\!] = 90°$

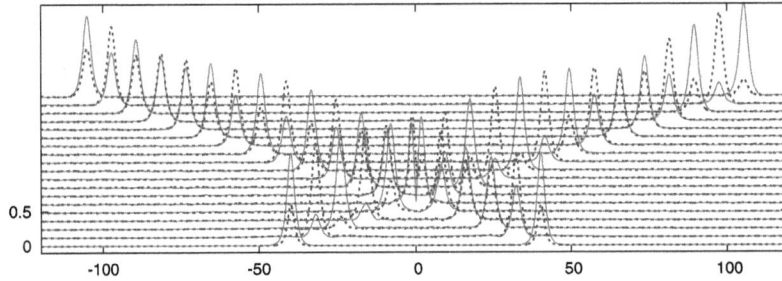

(c) $[\![\delta]\!] = 135°$

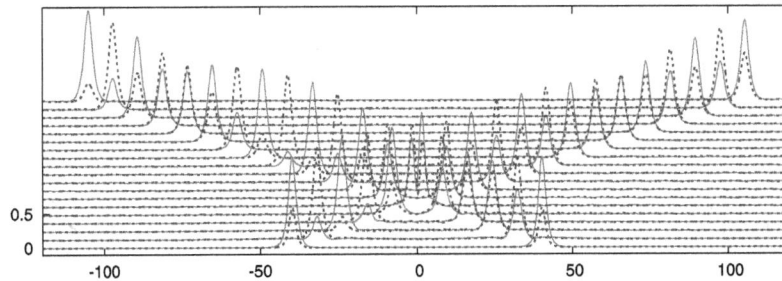

(d) $[\![\delta]\!] = 180°$

Figure 2.45. Head-on collision of QPs (ψ–solid line; ϕ–dashed line) with initial elliptic polarization for $\theta = 23°$ $44'$, $n_{l\,\psi} = n_{r\,\psi} = -1.1$, $n_{l\,\phi} = n_{r\,\phi} = -1.5$, $c_1 = -c_r = 1$, $\alpha_1 = 0.75$, $\Gamma = 0.175$, $X_1 = -40$, $X_r = 40$ and different initial phase differences.

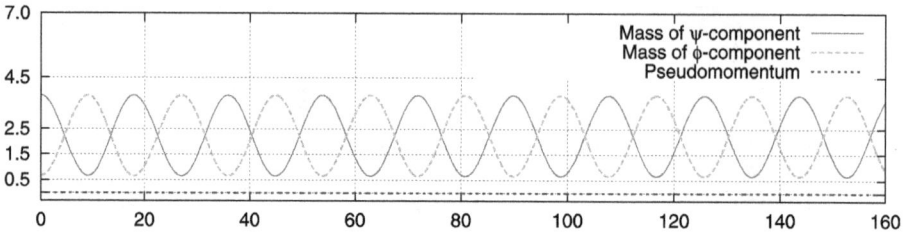

Figure 2.46. Mass and pseudomomentum dynamics for $[[\delta]] = 0°$, corresponding to figure 2.45.

polarization is general. The net conservation law is generalized as net breathing (oscillating) law, which conserves the total polarization within one time period. This result does not have an analytic analogue and enriches the phenomenology of CNLSE (figures 2.46 and 2.47). Among the three conservation laws about the mass, pseudomomentum and energy, we found one more net conservation law after an extensive set of numerical investigations.

Elliptically polarized solitons: takeover interaction
The pattern in takeover interaction is qualitatively the same but the required difference in the velocity magnitudes bears a bigger trajectory shift of the slower soliton and a slower trajectory shift of the bigger soliton (see [103]). Here, we consider this kind of interaction in order to implement an interaction of elliptically polarized solitons with different polarization angles. Also, such kinds of evolution and interaction do not possess symmetry. Our main goal is to check and confirm the conclusions about the net polarization realized in the previous subsection. We choose $n_{l\psi} \doteq n_{r\psi} = -1.1$, $n_{l\phi} = n_{r\phi} = -1.5$, $c_l = 1$, $c_r = 0.8$, $\alpha_1 = 0.75$, $\Gamma = 0.175$ and focus on the effects of $\overrightarrow{\delta}$. The above parametric set results in polarization angles $\theta_1 = 23° \, 45'$, $\theta_r = 25° \, 52'$ of the initial envelope configuration. The approximate initial condition is generated from the auxiliary bifurcation system. The solitons start with different elliptic polarizations and are an object of rotational polarization, due to nontrivial linear coupling Γ as well (figure 2.48). We find out that the initial phase differences of the components play an essential role on the magnitude of full energy of QPs. For example, when $[[\delta]] = 0° - E = 0.0673$; when $[[\delta]] = 90° - E = -0.976$; when $[[\delta]] = 180° - E = 0.657$. Just the opposite; the pseudomomentum is also conserved and does not depend on the initial phase difference. It is not trivial because of the nonsymmetry of the QP configuration and its net values P vary in the interval $(4.05, 4.08)$ (figure 2.49). The individual masses, however, breathe together with the individual (rotational) polarizations. Their amplitude and period are not influenced by the initial phase difference (figure 2.49) and are conserved within one full period of the breathing. The total mass is constant. Both the individual and total polarizations breathe and suffer a 'shock in polarization' when the QPs enter the collision. The polarization amplitude evidently depends on the initial phase difference (figure 2.50). The above quantities are conserved within one full period of the breathing. In all considered cases of evolution and interaction, the polarization angle of QPs can change independently of the collision due to the real linear coupling Γ.

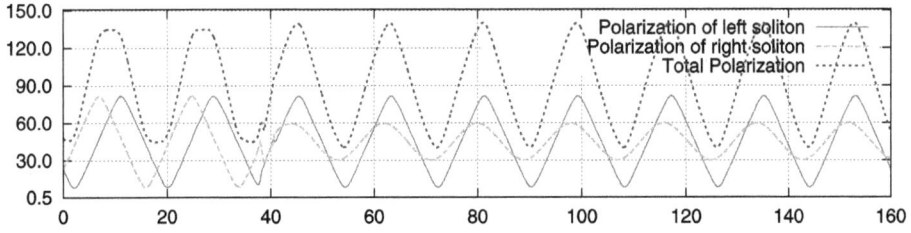

(a) $[\![\delta]\!] = 0°$

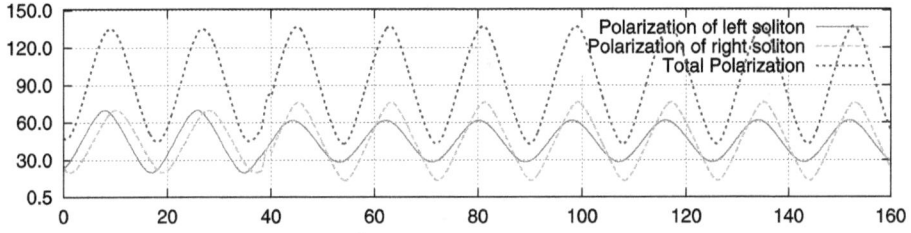

(b) $[\![\delta]\!] = 90°$

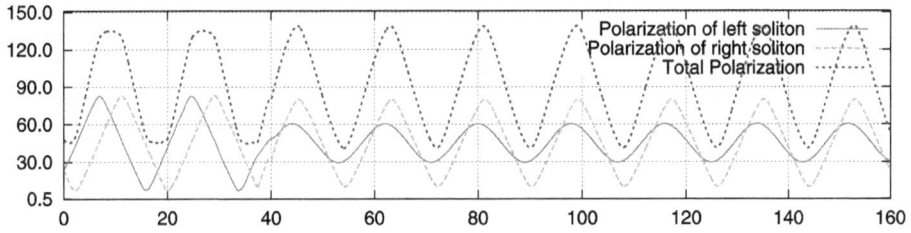

(c) $[\![\delta]\!] = 135°$

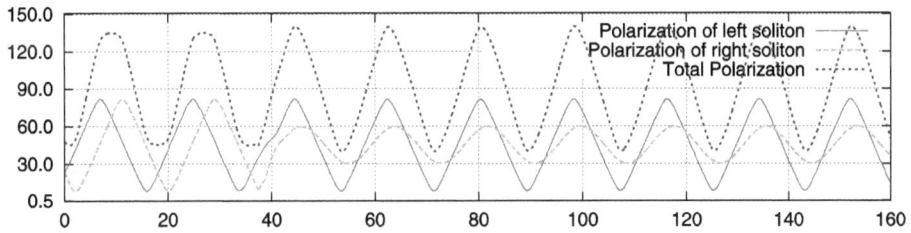

(d) $[\![\delta]\!] = 180°$

Figure 2.47. Individual and total polarization dynamics in dependence on the initial phase difference, corresponding to figure 2.45.

Conclusions

In this investigation, we aim to finalize the information about the dynamical properties of the total polarization influenced by the initial parameters and, in particular, by the initial polarization of the general type, as well as by nontrivial linear coupling Γ and the initial phase difference $[\![\delta]\!]$. For the case of general elliptic polarization, an exact initial condition concerning the shape of solitary envelopes is

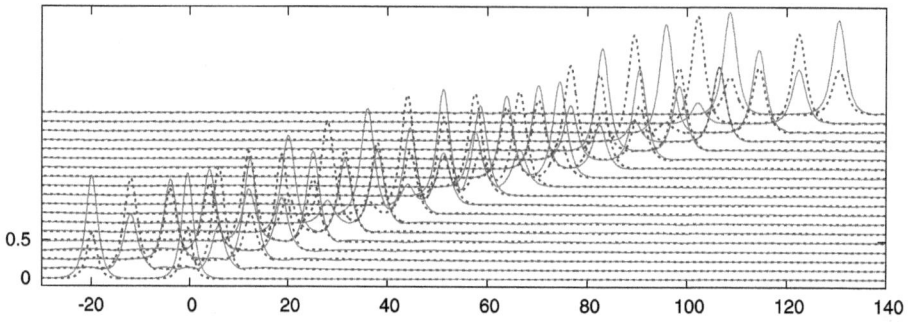

(a) $[\![\delta]\!] = 0°$

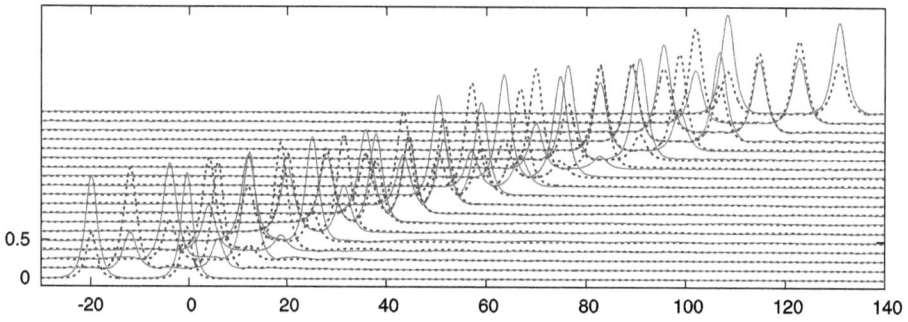

(b) $[\![\delta]\!] = 90°$

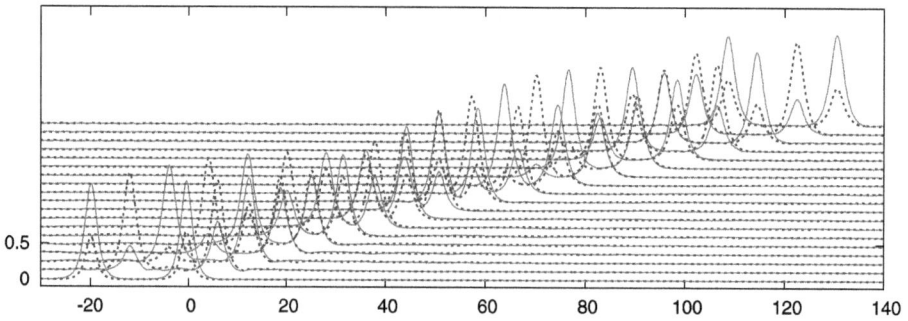

(c) $[\![\delta]\!] = 180$

Figure 2.48. Takeover collision of QPs (ψ–solid line; ϕ–dashed line) with initial elliptic polarization: polarization angles $\theta_l = 23° \, 45'$, $\theta_r = 25° \, 52'$, $c_l = 1$, $c_r = 0.8$, $n_{l\psi} = n_{r\psi} = -1.1$, $n_{l\phi} = n_{r\phi} = -1.5$, $\alpha_l = 0.75$, $\Gamma = 0.175$, $X_l = -20$, $X_r = 0$ and different initial phase differences.

not available. Solving an auxiliary bifurcation system of ordinary differential equations, we construct numerically approximate initial conditions for the full range of polarization angles between $0°$ and $90°$ and with a non-*sech* shape. Let us denote that the particular cases $\theta = 0°$; $90°$ (left and right linear polarization) when one has an exact *sech*-like initial condition were implemented in detail in [103]. We consider the two main interactions—head-on and takeover collisions of envelopes

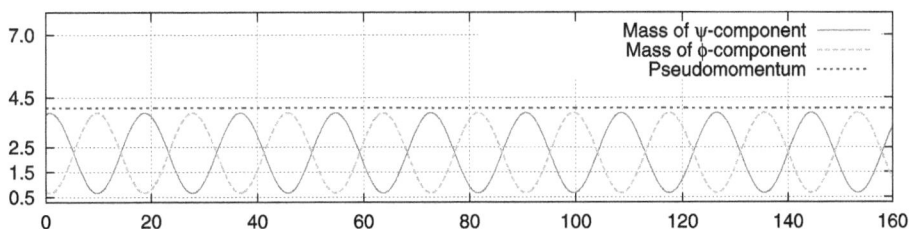

(a) $[\![\delta]\!] = 0°$

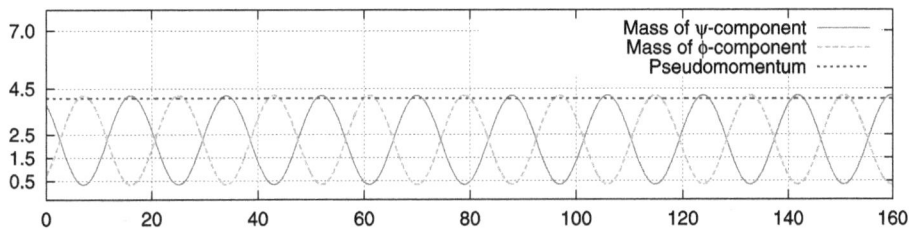

(b) $[\![\delta]\!] = 90°$

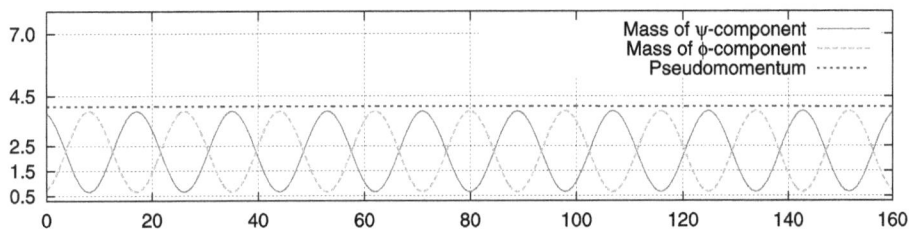

(c) $[\![\delta]\!] = 180°$

Figure 2.49. Mass and pseudomomentum dynamics for $[\![\delta]\!] = 0°$, corresponding to figure 2.48.

only when a real linear coupling Γ is present (the cross-modulation parameter $\alpha_2 = 0$). The adding of an imaginary part in Γ leads to the gain/blow-up of the solution accompanied by a violation of the conservation laws (see [103]) and we will not concern ourselves again with this property. The main conclusions can be summarized as follows

- The interaction perfectly conserves the pseudomomenta and energy. The pseudomomenta are not influenced by the initial phase difference $[\![\delta]\!]$, while the magnitude of the energy is strongly dependent on it. The variation is nonmonotonous.
- The total mass is also perfectly conserved but this is realized on the inner share and exchange of local masses of both components of QP periodically in time. This periodicity is governed by the real part of the linear coupling Γ and generates breathing of the masses, along with the breathing of the solitary envelopes. The mass breathing does not depend on the interaction and is present during the whole time of the evolution. The amplitude of mass oscillation depends on the initial phase difference.

(a) $[\![\delta]\!] = 0°$

(a) $[\![\delta]\!] = 90°$

(a) $[\![\delta]\!] = 180°$

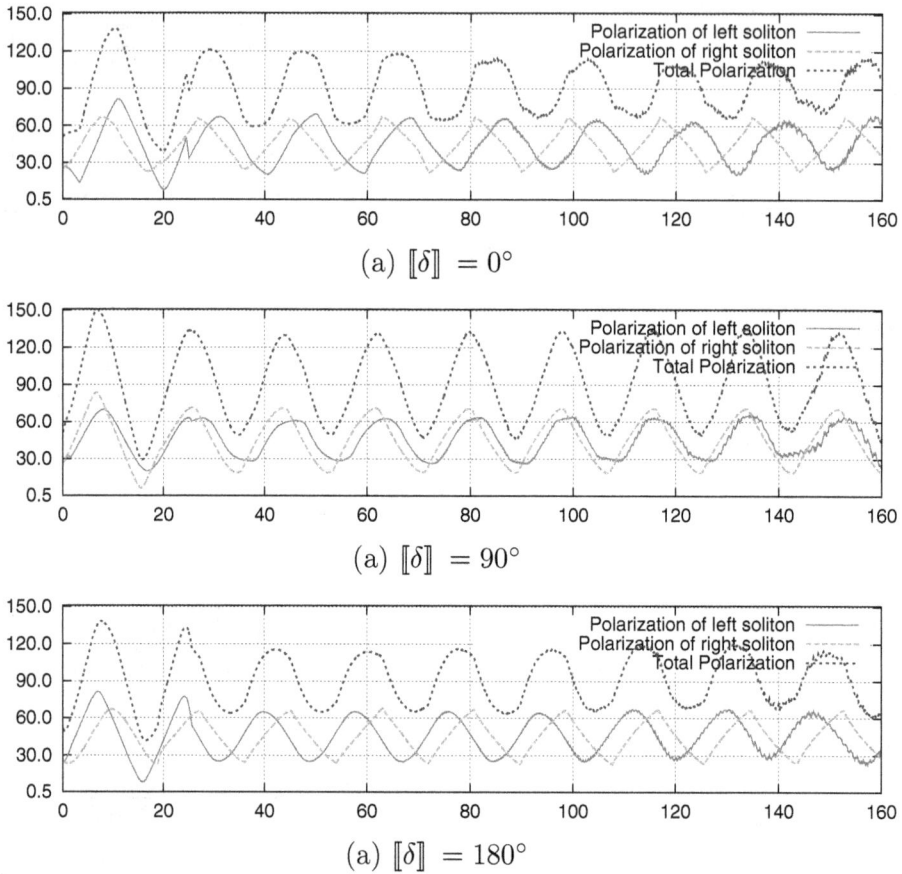

Figure 2.50. Individual and total polarization dynamics in dependence on initial phase difference, corresponding to figure 2.48.

- The individual polarizations breathe similarly to individual masses but opposite to them with a phase shift. This is the reason why the total rotational polarization is conserved within one period. In the place of interaction, the individual and total polarization angles suffer a discontinuity, keeping the rest of functions smooth. In other words, the total polarization also breathes and a bifurcation occurs during the cross-section of the interaction, and this is the main result of this investigation. The amplitudes of the individual polarizations are affected by the initial phase difference.
- For the case of rotational and arbitrary (elliptic) polarizations simultaneously, we found and generalized one more conservation law—for the net total polarization. So, the linearly CNLSE possess four conservation laws: for the total mass, for the pseudomomentum, for the energy, and for the magnitude of the total polarization. Let us emphasize that the last law does not have an analytical analogue and it unearths more intrinsic information for the propagating and interacting solitons as quasi-particles.

Another generalization of VNLSE is the system with different rates of the cross-modulation parameters. Though VNLSE remains nonintegrable, it has a bihamiltonian structure.

2.11 Vector nonlinear Schrödinger equation with different cross-modulation rates

Preliminary

In oceanography, freak waves (or rogue waves, extreme waves) are usually defined as waves whose height is more than twice the 'significant' wave height H_s (H_s is the mean of the largest third of waves in a wave record). The NLSE is a fundamental model for the slowly varying complex group envelope of surface waves over a deep water. The Peregrine soliton of the NLSE

$$\psi(x,\, t) = \left[1 - 4\frac{1 + 2it}{1 + 4x^2 + 4t^2}\right]\exp(it)$$

is often considered as a prototype model of the deterministic freak waves' generation [90]. The solution is with constant boundary conditions at $x \to \infty$ and has the property that at $x = 0$ and $t = 0$, the amplitude of the hump is 3 times higher than the uniform constant background. This suggests that an initially small 'hump' in a nearly monochromatic wave train may indeed evolve into a wave with a much larger amplitude. Thus, the nonlinear wave interaction is a mechanism of the creation of waves with significantly larger amplitude.

Hamiltonian of VNLSE with different rates of cross-modulation

Let us now consider the system

$$\begin{aligned}
i\psi_t &= \beta\psi_{xx} + [\alpha_1|\psi|^2 + (\alpha_1 + 2\alpha_{21})|\phi|^2]\psi, \\
i\phi_t &= \beta\phi_{xx} + [\alpha_1|\phi|^2 + (\alpha_1 + 2\alpha_{22})|\psi|^2]\phi.
\end{aligned} \tag{2.51}$$

Consider a transformation $\phi \to p\phi$ where $p^2 = \frac{\alpha_1 + 2\alpha_{22}}{\alpha_1 + 2\alpha_{21}}$. Then, the new Hamiltonian density should be

$$\mathcal{H} \overset{\text{def}}{=} \beta\left(|\psi_x|^2 + |\phi_x|^2\right) - \frac{1}{2}\alpha_1(|\psi|^4 + p^2|\phi|^4) - (\alpha_1 + 2\alpha_{22})(|\phi|^2|\psi|^2).$$

Consider a similar transformation of the original system $\psi \to q\psi$ with $q^2 = \frac{1}{p^2}$. Then, the Hamiltonian density looks like

$$\mathcal{H} \overset{\text{def}}{=} \beta\left(|\psi_x|^2 + |\phi_x|^2\right) - \frac{1}{2}\alpha_1(q^2|\psi|^4 + p^2|\phi|^4) - (\alpha_1 + 2\alpha_{21})(|\phi|^2|\psi|^2).$$

We have two Hamiltonians, which differ by a constant multiplier and therefore two completely isomorphic and equivalent Hamiltonian formulations of system (2.51).

Problem formulation: choice of initial conditions

We concern ourselves with the soliton solutions whose modulation amplitude is of *sech* form and which are localized envelopes on a propagating carrier wave. In the *sech*-case, the initial polarization can only be linear. Then, we assume that for each of the functions ψ, ϕ, the initial condition is of the form of a single propagating soliton, namely

$$\begin{Bmatrix} \psi(x,\,t) \\ \phi(x,\,t) \end{Bmatrix} = \begin{Bmatrix} A^{\psi} \\ A^{\phi} \end{Bmatrix} \mathrm{sech}[b(x - X - ct)]\exp\left\{ i\left[\frac{c}{2\beta}(x - X) - nt\right]\right\}.$$

$$b^2 = \frac{1}{\beta}\left(n + \frac{c^2}{4\beta}\right), \quad A = b\sqrt{\frac{2\beta}{\alpha_1}}, \quad u_c = \frac{2n\beta}{c},$$

(2.52)

where X is the spatial position (center of soliton), c is the phase speed, n is the carrier frequency, and b^{-1} is a measure of the support of the localized wave.

Numerical implementation

We investigate wave interactions modeled by two-component VNLSE (2.2) where each component describes an envelope of a separate wave packet. We show that such a nonlinear interaction causing polarization shock can lead to an amplification of the amplitude of one of the packets and hence provide a possible mechanism for the formation of freak waves. We use conservative fully implicit schemes (2.5) and (2.6) constructed earlier by us for investigation of the head-on collisions of solitary waves of VNLSE linearly polarized in the initial configuration.

Results and discussion

We elucidate numerically the role of different nonlinear couplings on their QP dynamics. We have uncover many other different scenarios of QP behavior upon collision including multiplying the soliton envelopes after the collision, dramatic change of amplitudes, and velocity shifts. In the first two figures, figures 2.51 and 2.52, one of the equations, (2.51), is like in the MS, i.e. with trivial cross modulation.

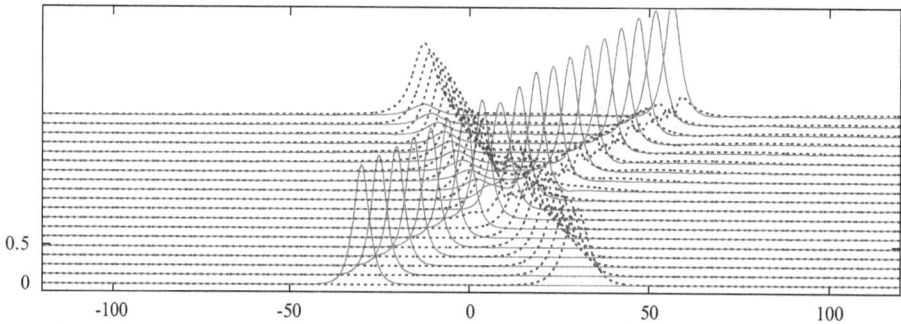

Figure 2.51. Mixed regime: only the first equation is like in the MS. $\alpha_{21} = 0$, $\alpha_{22} = 0.4$, $\alpha_1 = 0.25$, $\beta = 1$, $n_{\psi} = 0.2$, $n_{\phi} = 0.1$, $c_1 = 0.6$, $c_r = -0.3$.

The head-on collision causes a velocity shift, a jump transition from linear to elliptic polarization, and insignificant radiation and change of the envelope's amplitudes.

The next few plots (figures 2.53–2.56) cogently demonstrate the real effect of the different cross-modulation rates combined with carrier frequencies. The effects described above are complemented by a large number of newly born solitons and energetic excitations, abrupt change of the polarization angles and increase of the amplitudes, where the solitons become narrow.

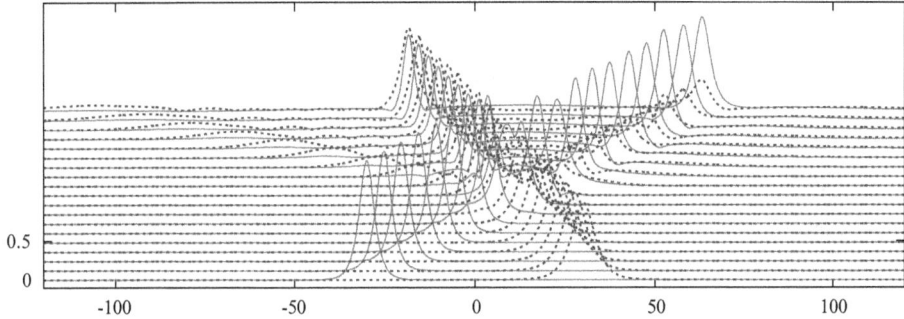

Figure 2.52. Mixed regime: only the second equation is like in the MS. $\alpha_{21} = 0.4$, $\alpha_{22} = 0$, $\alpha_1 = 0.25$, $\beta = 1$, $n_\psi = 0.2$, $n_\phi = 0.1$, $c_1 = 0.6$, $c_r = -0.3$.

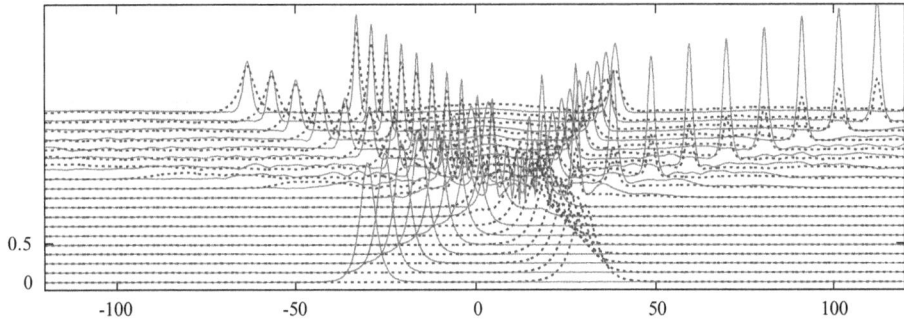

Figure 2.53. $\alpha_{21} = 4$, $\alpha_{22} = 0.4$, $\alpha_1 = 0.25$, $\beta = 1$, $n_\psi = 0.2$, $n_\phi = 0.1$, $c_1 = 0.6$, $c_r = -0.3$.

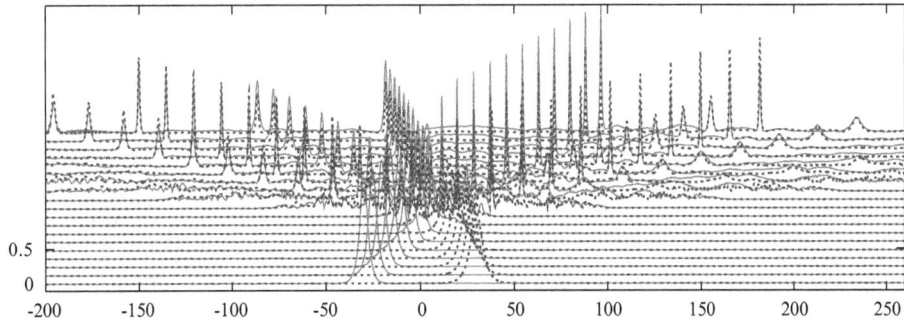

Figure 2.54. $\alpha_{21} = 4$, $\alpha_{22} = 8$, $\alpha_1 = 0.25$, $\beta = 1$, $n_\psi = 0.2$, $n_\phi = 0.1$, $c_1 = 0.6$, $c_r = -0.3$.

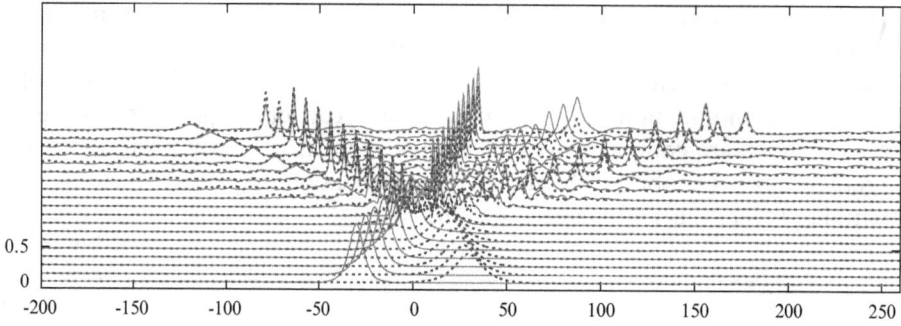

Figure 2.55. $\alpha_{21} = 4$, $\alpha_{22} = 8$, $\alpha_1 = 0.25$, $\beta = 1$, $n_\psi = n_\phi = 0$, $c_1 = 0.6$, $c_r = -0.3$.

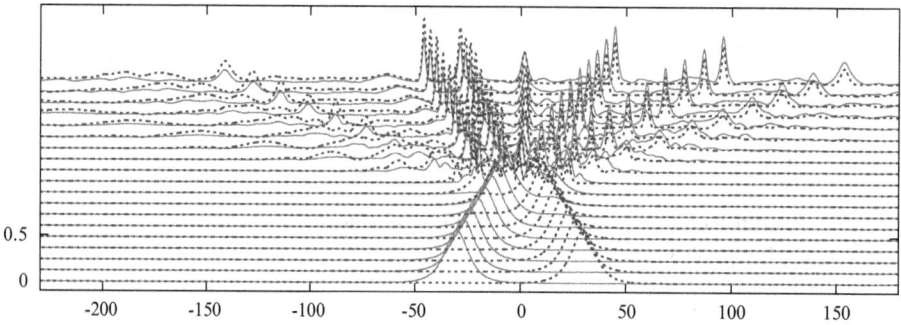

Figure 2.56. $\alpha_{21} = 4$, $\alpha_{22} = 8$, $\alpha_1 = 0.25$, $\beta = 1$, $n_\psi = n_\phi = 0$, $c_1 = -c_r = 0.4$.

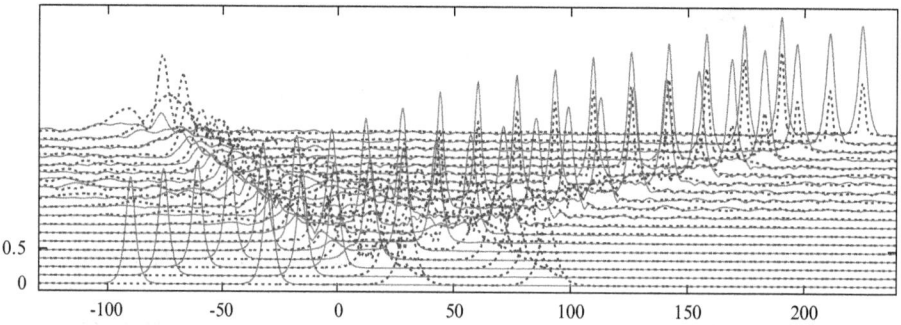

Figure 2.57. Mixed regime of two parallel pairs of two-component solitons: only the second equation is like Manakov. $\alpha_{21} = 0.4$, $\alpha_{22} = 0$, $\alpha_1 = 0.25$, $\beta = 1$, $n_\psi = 0.2$, $n_\phi = 0.1$, $c_1 = 0.6$, $c_r = -0.3$.

These simulations clearly show why the colliding solitons have to be considered as QPs. They produce a possible way to simulate freak waves by varying the cross modulations and carrier frequencies of the colliding soliton envelopes. The last graph, figure 2.57, presents a head-on collision of two parallel propagating pairs of two-component solitons with different cross modulations. The second equation in (2.51) is like Manakov, i.e. $\alpha_{22} = 0$.

VNLSE can be developed by increasing either the number of soliton components or the number of initial envelopes. Such a kind of generalization of the finite-difference approach with a conservative scheme and an internal iteration is useful and immediately implemented for Bose–Einstein condensates.

Our next considerations will be devoted to the perturbed by external potentials VNLSE in the absence of cross modulation. The system in this case is called MS and we have discussed it repeatedly. Opposite to QP dynamics and behavior of its soliton solutions, we will accent the dynamics and asymptotic behavior of soliton trains in the so-called adiabatic approximation. In this case, the original (perturbed) MS can be replaced with a dynamical system with respect to the soliton parameters— velocities, amplitudes, phases and polarization vectors.

2.12 Asymptotic behavior of Manakov solitons

We consider the asymptotic behavior of the soliton solutions of MS perturbed by external potentials. It has already been established that its multisoliton interactions in the adiabatic approximation can be modeled by the complex Toda chain (CTC). The fact that the CTC is a completely integrable system enables us to determine the asymptotic behavior of the multisoliton trains. In the present study, we accent the 3-soliton initial configurations perturbed by *sech*-like external potentials and compare the numerical predictions of the MS (as a particular case of VNLSE with a trivial cross-modulation) and the perturbed CTC in different regimes. The results of the conducted analysis show that the perturbed CTC can reliably predict the long-time evolution of the MS.

Introduction

The GP equation and its multicomponent generalizations are important tools for analyzing and studying the dynamics of the Bose–Einstein condensates (BEC); see the monographs [50, 62, 85] and the numerous references therein, among which we mention [14, 16, 33, 41, 56, 57, 63, 65, 71, 77, 81, 107, 108]. In the three-dimensional case, these equations can be analyzed solely by numerical methods. If we assume that BEC is quasi-one-dimensional, then the GP equations mentioned above may be reduced to the NLSE perturbed by the external potential $V(x)$

$$i\psi_t + \frac{1}{2}\psi_{xx} + |\psi|^2\psi(x, t) = V(x)\psi(x, t),\tag{2.53}$$

or to its vector generalizations (VNLSE)

$$i\vec{u}_t + \frac{1}{2}\vec{u}_{xx} + (\vec{u}^\dagger, \vec{u})\vec{u}(x, t) = V(x)\vec{u}(x, t).\tag{2.54}$$

The Manakov model [72] is a two-component VNLSE with $\vec{u} = (u_1, u_2)^T$ and $V(x) = 0$ (for more details, see [6, 48]).

The analytical approach to the N-soliton interactions was proposed by Zakharov and Shabat [80, 114] for the scalar NLSE; for a VNLSE, see [64]. They treated the case of the exact N-soliton solution where all solitons had different velocities. They calculated the asymptotics of the N-soliton solution for $t \to \pm\infty$ and proved that both asymptotics are the sums of N one-soliton solutions with the same sets of amplitudes and velocities. The effects of the interaction were shifts in the relative center of masses and phases of the solitons. The same approach, however, is not applicable to the Manakov model, because the asymptotics of the soliton solution for $t \to \pm\infty$ do not commute.

The present investigation is an extension of [35, 42] where the main result was that the N-soliton interactions in the adiabatic approximation for the Manakov model can also be modeled by the CTC [38, 39, 46, 47]. More specifically, here we continue our analysis of the effects of external potentials on the soliton interactions. While in [33, 35, 42], we studied the effects of periodic, harmonic and anharmonic potentials, here we consider potential wells and humps of the form:

$$V(x) = \sum_{s=1}^{A} c_s V_s(x, y_s), \qquad V_s(x, y_s) = \frac{1}{\cosh^2(2\nu_0 x - y_s)}. \qquad (2.55)$$

If c_s is negative (positive), $V_s(x)$ is a well (hump) with a width of 1.7 at half-height/depth. Adjusting one or more terms in (2.55) with different c_s and y_s, we can describe wells and/or humps with different widths/depths and positions.

We in fact prove the hypothesis in [32] and extend the results in [30, 32–35, 40, 70, 89, 113] concerning the model of soliton interactions of VNLSE (2.54) in adiabatic approximation.

The corresponding vector N-soliton train is a solution of (2.54) determined by the initial condition:

$$\vec{u}(x, t = 0) = \sum_{k=1}^{N} \vec{u}_k(x, t = 0), \quad \vec{u}_k(x, t) = u_k(x, t)\vec{n}_k, \quad u_k(x, t) = \frac{2\nu_k e^{i\phi_k}}{\cosh(z_k)} \qquad (2.56)$$

with

$$z_k = 2\nu_k(x - \xi_k(t)), \qquad \xi_k(t) = 2\mu_k t + \xi_{k,0},$$

$$\phi_k = \frac{\mu_k}{\nu_k} z_k + \delta_k(t), \qquad \delta_k(t) = 2(\mu_k^2 + \nu_k^2)t + \delta_{k,0}, \qquad (2.57)$$

where the s-component polarization vector $\vec{n}_k = (n_{k,1} e^{i\beta_{k,1}}, n_{k,2} e^{i\beta_{k,2}}, \ldots, n_{k,s} e^{i\beta_{k,s}})^T$ is normalized by the conditions

$$\langle \vec{n}_k^\dagger, \vec{n}_k \rangle \equiv \sum_{p=1}^{s} n_{k,p}^2 = 1, \; \sum_{p=1}^{s} \beta_{k;s} = 0. \qquad (2.58)$$

The adiabatic approximation holds true if the soliton parameters satisfy [61]:

$$|\nu_k - \nu_0| \ll \nu_0, \qquad |\mu_k - \mu_0| \ll \mu_0, \qquad |\nu_k - \nu_0||\xi_{k+1,0} - \xi_{k,0}| \gg 1, \qquad (2.59)$$

for all k, where $\nu_0 = \frac{1}{N}\sum_{k=1}^{N} \nu_k$ and $\mu_0 = \frac{1}{N}\sum_{k=1}^{N} \mu_k$ are the average amplitude and velocity, respectively. In fact, we have two different scales:

$$|\nu_k - \nu_0| \simeq \varepsilon_0^{1/2}, \qquad |\mu_k - \mu_0| \simeq \varepsilon_0^{1/2}, \qquad |\xi_{k+1,0} - \xi_{k,0}| \simeq \varepsilon_0^{-1}.$$

We remind you that the basic idea of the adiabatic approximation is to derive a dynamical system for the soliton parameters, which would describe their interaction.

This idea was initiated by Karpman and Solov'ev [61] and modified by Anderson and Lisak [13]. Later, this idea was generalized to N-soliton interactions [38, 39, 46, 47] and the corresponding dynamical system for the 4N-soliton parameters was identified as a N-site CTC. The fact that the CTC (just like its real counterpart—the Toda chain) is completely integrable gives additional possibilities. A detailed comparative analysis between the solutions of the RTC and CTC [36, 37] shows that the CTC allows for a variety of asymptotic regimes; see the next section below. More precisely, knowing the initial soliton parameters, one can effectively predict the asymptotic regime of the soliton train. Another possible use of the same fact is that one can describe the sets of soliton parameters responsible for each of the asymptotic regimes. Another important advantage of the adiabatic approach consists of the fact that one may consider the effects of various perturbations on the soliton interactions [46, 61].

The next step was to extend this approach to treat the soliton interactions of the Manakov solitons. More precisely, using the method of Anderson and Lisak [13], we derive a generalized version of the CTC (see equations (2.68) and (2.69) below) as a model describing the behavior of the N-soliton trains of the VNLSE (2.54) [30, 32, 33, 35, 40]. This generalized CTC includes dependence on the polarization vectors $\vec{n}_k$. It allows us to analyze how the changes of the polarization vectors influence the soliton interactions. Besides, the generalized CTC is also integrable with the consequence that one can predict the asymptotic regimes of the Manakov solitons and can describe the sets of soliton parameters that are responsible for each of the asymptotic regimes. Of course, just like for the scalar case, one can also analyze the effects of the various perturbations on the soliton interactions.

Based on the above, we aim to:

1. Outline how the variational approach developed in [13] can be used to derive the perturbed CTC (PCTC) model [30, 35] for *sech*-type external potentials. We also remind the reader about the asymptotic regimes of the soliton trains predicted by the CTC [38, 39].

2. Briefly treat the N-soliton interactions of the MS without external potential. Obviously, in order to determine the N-soliton train for the MS, along with the usual sets of solitons parameters ν_k, μ_k, ξ_k and δ_k, we also need the set of polarization vectors $\vec{n}_s$.

3. Derive the effects of the external potentials on the soliton interactions. This is a PCTC for generic potentials of the form (2.55).

4. Compare the numerical solutions of the perturbed VNLSE (2.54) with the predictions of the PCTC model. To this end, we solve the VNLSE numerically by using an implicit scheme of Crank-Nicolson type in complex arithmetic, equations (2.5) and (2.6). The concept of the internal iterations is applied (see [23]) in order to ensure the implementation of the conservation laws on difference level within the round-off error of the calculations [99, 100, 105]. The solutions of the relevant PCTC have been obtained using Maple. Knowing the numeric solution $\vec{u}$ of the perturbed VNLSE, we calculate the maxima of $(\vec{u}^{\dagger}, \vec{u})$, compare them with the numerical solutions

for $\xi_k(t)$ of the PCTC and plot the models trajectories predicted by both for each of the solitons. Thus, we are able to analyze the effects of the external potentials on the soliton interactions.

Preliminaries

Here, we briefly remind you of the derivation of the CTC as a model, describing the N-soliton interactions VNLSEs using the variational approach [13].

Derivation of the CTC as a model for the soliton interaction of perturbed VNLSEs
The perturbed VNLSE (2.54) allows Hamiltonian formulations with the Poisson brackets

$$\{\vec{u}_j(x,\,t),\,\vec{u}_k^*(y,\,t)\} = \delta_{jk}\delta(x-y) \tag{2.60}$$

and the Hamiltonian

$$H = \int_{-\infty}^{\infty} dx \left[\frac{1}{2}\langle\vec{u}_x,\,\vec{u}_x\rangle - \frac{1}{2}\langle\vec{u}^\dagger,\,\vec{u}\rangle^2 + V(x)\langle\vec{u}^\dagger,\,\vec{u}\rangle\right]. \tag{2.61}$$

It also admits a Lagrangian of the form:

$$\mathcal{L} = \int_{-\infty}^{\infty} dt\,\frac{i}{2}[\langle\vec{u},\,\vec{u}_t\rangle - \langle\vec{u}_t,\,\vec{u}\rangle] - H. \tag{2.62}$$

In what follows, we will analyze the large time behavior of the N-soliton train determined as the solution of the VNLSE by the initial conditions (2.56) and (2.57).

The idea of the variational approach of [13] is to insert the ansatz (2.56) into the Lagrangian, perform the integration over x and retain only terms of the orders of $\varepsilon_0^{1/2}$ and ε_0. The first obvious observation is that only the nearest neighbor solitons will contribute such terms and

$$\mathcal{L} = \sum_{k=1}^{N}\mathcal{L}_k + \sum_{k=1}^{N}\sum_{n=k\pm1}\mathcal{L}_{kn}. \tag{2.63}$$

Here, $\mathcal{L}_k$ corresponds to the terms involving only the kth soliton (see [35]):

$$\begin{aligned}
\mathcal{L}_k = 4\nu_k &\left(\frac{i}{2}(\langle\vec{n}_k,\,\vec{n}_{k,t}\rangle - \langle\vec{n}_{k,t},\,\vec{n}_k\rangle) + 2\mu_k\frac{d\xi_k}{dt} - \frac{d\delta_k}{dt} - 2\mu_k^2 + \frac{2\nu_k^2}{3}\right) \\
&- \int_{x=-\infty}^{\infty} V(x)\langle\vec{u}_k^\dagger,\,\vec{u}_k\rangle.
\end{aligned} \tag{2.64}$$

Finally, the terms describing soliton–soliton interactions are given by:

$$\begin{aligned}
&\mathcal{L}_{kn} = 16\nu_0^3 e^{-\Delta_{kn}}(R_{kn} + R_{kn}^*) + \mathcal{O}(\epsilon^{3/2}), \\
&R_{kn} = e^{i(\tilde{\delta}_n - \tilde{\delta}_k)}\langle\vec{n}_k,\,\vec{n}_n\rangle, \qquad \tilde{\delta}_k = \delta_k - 2\mu_0\xi_k, \quad \Delta_{kn} = 2s_{kn}\nu_0(\xi_k - \xi_n),
\end{aligned} \tag{2.65}$$

where $s_{k,k+1} = 1$ and $s_{k,k-1} = -1$.

The next step is to consider $\mathcal{L}$ (2.63) as a Lagrangian of the dynamical system, describing the motion of the N-soliton train and providing the equations of motion for the $(2s + 2)N$ ($6N$ for the Manakov case) soliton parameters.

Let us first consider the unperturbed case, i.e. $V(x) = 0$. Deriving the dynamical system, we get terms of three different orders of magnitude: (i) terms of order $\Delta_{kn}^2 \exp(-\Delta_{kn})$; (ii) terms of order $\Delta_{kn} \exp(-\Delta_{kn})$ and (iii) terms of order $\exp(-\Delta_{kn})$. However, the terms of types (i) and (ii) are multiplied by factors that are of the order of $\exp(-\Delta_{kn})$, due to the evolution equations for the soliton parameters. Finally, we arrive at the following set of dynamical equations for the soliton parameters:

$$\frac{d\xi_k}{dt} = 2\mu_k, \qquad\qquad \frac{d\delta_k}{dt} = 2\mu_k^2 + 2\nu_k^2,$$

$$\frac{d\nu_k}{dt} = 8\nu_0^3 \sum_n e^{-\Delta_{kn}} i(R_{kn} - R_{kn}^*), \qquad \frac{d\mu_k}{dt} = -8\nu_0^3 \sum_n e^{-\Delta_{kn}}(R_{kn} + R_{kn}^*). \tag{2.66}$$

In addition, we also obtain a system of equations for the evolution of the polarization vectors:

$$\frac{d\vec{n}_k}{dt} = 4\nu_0^2 i \sum_{n=k\pm 1} e^{-\Delta_{kn}} [e^{i(\tilde{\delta}_n - \tilde{\delta}_k)}\vec{n}_n - R_{kn}\vec{n}_k + e^{i(\tilde{\delta}_n - \tilde{\delta}_k)}\vec{n}_n + R_{kn}^*\vec{n}_k] + C_k\vec{n}_k \tag{2.67}$$

where the constants C_k are fixed up by the constraints on the polarization vectors. Indeed, from $\langle \vec{n}_k^\dagger, \vec{n}_k \rangle = 1$ for all t, one finds that $C_k + C_k^* = 0$, i.e. the constants C_k are purely imaginary. Let us now assume that $C_k = i\theta_k$. Then, from equations (2.58) and (2.67), we find that $\beta_{k,s}$ becomes time-dependent and up to terms of the order of ϵ evolves linearly with time: $\beta_{k,s}(t) = \beta_{k,s}(0) + \theta_k t$. But such an evolution is only compatible with the second normalization condition in (2.58) if $\theta_k = 0$; therefore $C_k = 0$. Thus, from equations (2.66), we get:

$$\frac{d(\mu_k + i\nu_k)}{dt} = 4\nu_0[\langle \vec{n}_k, \vec{n}_{k-1} \rangle e^{q_k - q_{k-1}} - \langle \vec{n}_{k+1}, \vec{n}_k \rangle e^{q_{k+1} - q_k}], \tag{2.68}$$

where

$$q_k = -2\nu_0 \xi_k + k \ln 4\nu_0^2 - i(\delta_k + \delta_0 + k\pi - 2\mu_0 \xi_k),$$

$$\nu_0 = \frac{1}{N} \sum_{s=1}^N \nu_s, \qquad \mu_0 = \frac{1}{N} \sum_{s=1}^N \mu_s, \qquad \delta_0 = \frac{1}{N} \sum_{s=1}^N \delta_s. \tag{2.69}$$

Besides, from (2.66) and (2.69), there follows (see [46]):

$$\frac{dq_k}{dt} = -4\nu_0 (\mu_k + i\nu_k) \tag{2.70}$$

and

$$\frac{d^2 q_k}{dt^2} = 16\nu_0^2[\langle \vec{n}_{k+1}, \vec{n}_k \rangle e^{q_{k+1} - q_k} - \langle \vec{n}_k, \vec{n}_{k-1} \rangle e^{q_k - q_{k-1}}], \tag{2.71}$$

which prove the statement in [32]. Equation (2.71), combined with the system of equations for the polarization vectors (2.67), provides the proper generalization of the CTC model for the MS.

The equations for the polarization vectors are nonlinear. So, the whole system of equations for q_k and $\vec{n}_k$ seems to be rather complicated and nonintegrable, even for the unperturbed MS. However, all terms in the right hand sides of the evolution equations for $\vec{n}_k$ are of the order of ϵ. This allows us to neglect the evolution of $\vec{n}_k$ and to approximate them with their initial values. As a result, we obtain that the N-soliton interactions for the VNLSE in the adiabatic approximation are modeled by the CTC.

It is easy to see that, if all $\langle \vec{n}_{k+1}^{\dagger}, \vec{n}_k \rangle = \text{const} \neq 0$, then the CTC (2.71) is a completely integrable dynamical system, just like the RTC.

Note also that the CTC models the soliton interactions for the VNLSE with *any number of components*. The effect of the polarization vectors on the interaction comes into CTC only through the scalar products $m_{0k} = \langle \vec{n}_{k+1}, \vec{n}_k \rangle$. It is well known that a gauge transformation $\vec{u} \to g_0 \vec{u}$ with any constant unitary matrix g_0 leaves the VNLSE, equation (2.54), invariant. Such a transformation will change all polarization vectors simultaneously $\vec{n}_k \to g_0 \vec{n}_k$ but preserves their scalar products, and so will not influence the soliton interaction. Obviously, our CTC model is invariant under such transformations. Due to the above arguments, our choice of the initial values of $\vec{n}_{0k}$ can be changed into $g_0 \vec{n}_{0k}$ with no effect on the interaction. That is why we specify only the scalar products m_{0k} for our runs, which we have chosen to be real.

The effects of the sech-*like potentials on CTC*
Now we consider the effects of the external potentials of the form (2.55). To this end, we have to calculate the integrals in the right-hand side of equation (2.64) and see how they would change the right-hand sides of equations (2.66).

As is clear from above, we have to replace $\mathcal{L}_k$ in equation (2.64) by

$$\mathcal{L}_{k,\,\text{pert}} = \mathcal{L}_k - 2\nu_k \int_{-\infty}^{\infty} \frac{dx\ V(x)}{\cosh^2(z_k)} \tag{2.72}$$

while $\mathcal{L}_{kn}$ remains unchanged. Thus, we obtain the following PCTC system:

$$\frac{d\lambda_k}{dt} = -4\nu_0\left(e^{q_{k+1}-q_k}(\vec{n}_{k+1}^{\dagger}, \vec{n}_k) - e^{q_k-q_{k-1}}(\vec{n}_k^{\dagger}, \vec{n}_{k-1})\right) + M_k + iN_k,$$

$$\frac{dq_k}{dt} = -4\nu_0\lambda_k + 2i(\mu_0 + i\nu_0)\Xi_k - iX_k, \qquad \frac{d\vec{n}_k}{dt} = O(\epsilon), \tag{2.73}$$

where $\lambda_k = \mu_k + i\nu_k$, $X_k = 2\mu_k \Xi_k + D_k$ and

$$N_k = -\frac{1}{2} \int_{-\infty}^{\infty} \frac{dz_k}{\cosh z_k}\ \Im\left(V(y_k)u_k e^{-i\phi_k}\right), \quad M_k = \frac{1}{2} \int_{-\infty}^{\infty} \frac{dz_k \sinh z_k}{\cosh^2 z_k}\ \Re\left(V(y_k)u_k e^{-i\phi_k}\right),$$

$$\Xi_k = -\frac{1}{4\nu_k^2} \int_{-\infty}^{\infty} \frac{dz_k\ z_k}{\cosh z_k}\ \Im\left(V(y_k)u_k e^{-i\phi_k}\right),$$

$$D_k = \frac{1}{2\nu_k} \int_{-\infty}^{\infty} \frac{dz_k\,(1 - z_k \tanh z_k)}{\cosh z_k}\ \Re\left(V(y_k)u_k e^{-i\phi_k}\right),$$

where $y_k = z_k/(2\nu_0) + \xi_k$.

As a result for our specific choice of $V(x)$, we get:

$$M_k = \sum_s 2c_s\nu_k P(\Delta_{k,s}), \qquad N_k = 0, \qquad \Xi_k = 0, \qquad D_k = \sum_s c_s R(\Delta_{k,s}), \quad (2.74)$$

where $\Delta_{k,s} = 2\nu_0\xi_k - y_s$ and the integrals describing the interaction of the solitons with the potential (see figure 2.58 (left panel))

$$P(\Delta) = \frac{\Delta + 2\Delta\cosh^2(\Delta) - 3\sinh(\Delta)\cosh(\Delta)}{\sinh^4(\Delta)},$$

$$R(\Delta) = \frac{6\Delta\sinh(\Delta)\cosh(\Delta) - (2\Delta^2 + 3)\sinh^2(\Delta) - 3\Delta^2}{2\sinh^4(\Delta)}. \qquad (2.75)$$

The details of deriving the integrals are given in [44].

The corrections to N_k and P_k, coming from the terms linear in u depend only on the parameters of the kth soliton; i.e. they are 'local' in k.

CTC and the asymptotic regimes of N-soliton trains

The fact that the N-soliton trains for the scalar NLSE are modeled by an integrable model, CTC allows one to to predict their asymptotic behavior. The method to do so was based on the exact integrability of the CTC [39] and on its Lax representation.

Here, we shall show that similar results hold true also for the CTC (2.66) modeling the soliton trains of the MS. Indeed, following Moser [78], we introduce the Lax pair

$$\dot{L} = [B, L], \qquad (2.76)$$

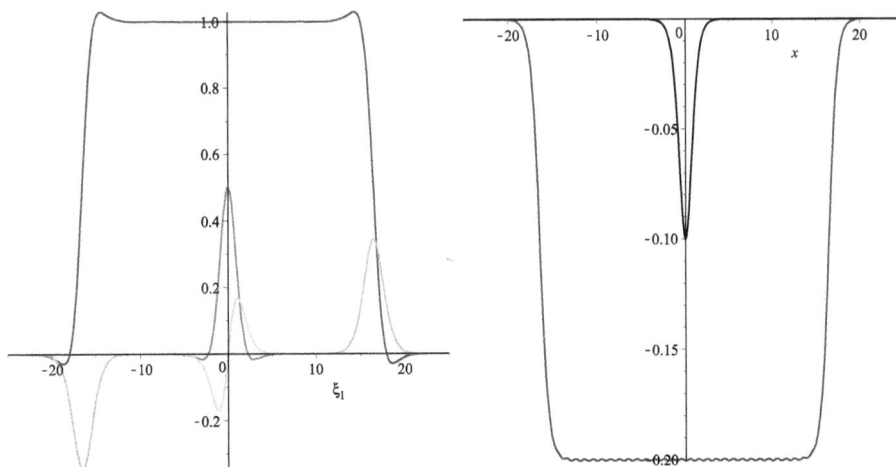

Figure 2.58. Graphs of P and R functions: for a single sech-potential centered at the origin in *cyan* and *red*; and for the superposed potential at the neighboring panel in *green* and *brown* (left); single *sech*-potential in *black* vs superposed external potential $V(x) = \sum_{s=0}^{32} c_s \operatorname{sech}^2(x - x_s)$, $c_s = -10^{-1}$, $x_s = -16 + sh$, $h = 1$, $s = 0, ..., 32$ in *blue*. The superposed potential forms a well (right).

where

$$L = \sum_{k=1}^{N}(b_k E_{kk} + a_k(E_{k,\,k+1} + E_{k-1,\,k})),$$

$$B = \sum_{k=1}^{N} a_k(E_{k,\,k+1} - E_{k-1,\,k}).$$

(2.77)

Here, the matrices $(E_{kn})_{pq} = \delta_{kp}\delta_{nq}$ and $E_{kn} = 0$ whenever one of the indices becomes 0 or $N + 1$; the other notations in (2.76) are as follows:

$$a_k = \frac{1}{2}\sqrt{\langle \vec{n}_{k+1},\, \vec{n}_k \rangle}\, e^{(q_{k+1}-q_k)/2}, \qquad b_k = \frac{1}{2}(\mu_k + i\nu_k).$$

(2.78)

One can check that the compatibility of condition equation (2.76) with L and B as in (2.77) is equivalent to the unperturbed CTC (2.66).

The first consequence of the Lax representation is that the CTC has N complex-valued integrals of motion provided by the eigenvalues of L, which we denote by $\zeta_k = \kappa_k + i\eta_k$, $k = 1, ..., N$. Indeed, the Lax equation means that the evolution of L is isospectral, i.e. $d\,\zeta_k/dt = 0$.

Another important consequence from the results of Moser [78] is that for the RTC, one can write down explicitly its solutions in terms of the scattering data, which consist of $\{\zeta_k, r_k\}_{k=1}^{N}$ where r_k are the first components of the properly normalized eigenvectors of L_0 [78, 98]. For the RTC, both $\zeta_k = \kappa_k$ and r_k are real; besides this, all ζ_k are different. Next, Moser calculated the asymptotics of these solutions for $t \to \pm\infty$ and showed that κ_k determine the asymptotic velocities of the particles.

The formulae derived by Moser can easily be extended to the complex case [78]. The important difference is that all important ingredients, such as eigenvalues ζ_k and first components of the eigenvectors of L normalized to 1 now, become complex valued. In addition, the important asset of L for the RTC, namely that all eigenvalues are real and different, is also lost. However, the asymptotics of the solutions for $t \to \pm\infty$ can be calculated with the result:

$$q_k(t) = -2\nu_0\zeta_k t - B_k + \mathcal{O}(e^{-Dt}),$$

(2.79)

where D in (2.79) is some real positive constant, which is estimated by the minimal difference between the asymptotic velocities. Equating the real parts in equation (2.79), we obtain:

$$\lim_{t \to \infty}(\xi_k + 2\kappa_k t) = \text{const}$$

(2.80)

which means that the real parts κ_k of the eigenvalues of L determine the asymptotic velocities for the CTC. This fact will be used to classify the regimes of asymptotic behavior.

Let us also point out the important differences between RTC and CTC, namely:
D1) While for RTC q_k, r_k and ζ_k are all real, for CTC they generically take complex values, e.g. $\zeta_k = \kappa_k + i\eta_k$;
D2) While for RTC $\zeta_k \neq \zeta_j$ for $k \neq j$, for CTC no such restriction holds.

As a consequence, we find that the only possible asymptotic behavior in the RTC is the asymptotically free motion of the solitons. For CTC, it is κ_k that determines the asymptotic velocity of the kth soliton. For simplicity and without loss of generality, we assume that: tr $L_0 = 0$; $\zeta_k \neq \zeta_j$ for $k \neq j$; and $\kappa_1 \leqslant \kappa_2 \leqslant \ldots \leqslant \kappa_N$. Then, we have:

AFR The asymptotically free regime takes place if $\kappa_k \neq \kappa_j$ for $k \neq j$, i.e. the asymptotic velocities are all different. Then, we have asymptotically separating free solitons, see also [38, 46];

BSR The bound state regime takes place for $\kappa_1 = \kappa_2 = \ldots = \kappa_N = 0$, i.e. all N solitons move with the same mean asymptotic velocity, and form a 'bound state'. The key question now will be the nature of the internal motions in such a bound state: is it quasi-equidistant or not?

MAR A variety of intermediate situations, or mixed asymptotic regimes happen when one group (or several groups) of particles move with the same mean asymptotic velocity; then they would form one (or several) bound state(s) and the rest of the particles will have free asymptotic motion.

Obviously, the regimes BSR and MAR, as well as the degenerate and singular cases, which we do not consider here, have no analogies in the RTC and physically are qualitatively different from AFR.

The perturbed CTC, taking into account the effects of the *sech*-like potentials, to the best of our knowledge is not integrable and does not allow Lax representation. Therefore, we are applying numerical methods to solve it.

Comparison between the PCTC model and Manakov soliton interactions

We will compare how well the PCTC model derived above predicts the soliton interactions as solutions of the Manakov model with the external potentials of kind (2.55). Before that, let us remind you of the well known result (see [38, 39, 46]) that the 3-soliton systems allow for three types of dynamical regimes for large times, namely

AFR Asymptotically free regime when all three solitons move away with different velocities. This regime takes place if the initial amplitudes are given by equation (2.83) with [38]:

$$\Delta v < v_{cr} = 2\sqrt{2\cos(\theta_1 - \theta_2)}\, v_0 \exp(-v_0 r_0) \tag{2.81}$$

and the phases are as in (2.84a). For our configuration with $r_0 = 8$, we have $v_{cr} = 0.0246$. Such an asymptotic regime is shown in the left panel of figure 2.59;

MAR Mixed asymptotic regime, when two of the solitons form a bound state and the third soliton goes away from them with a different velocity; such

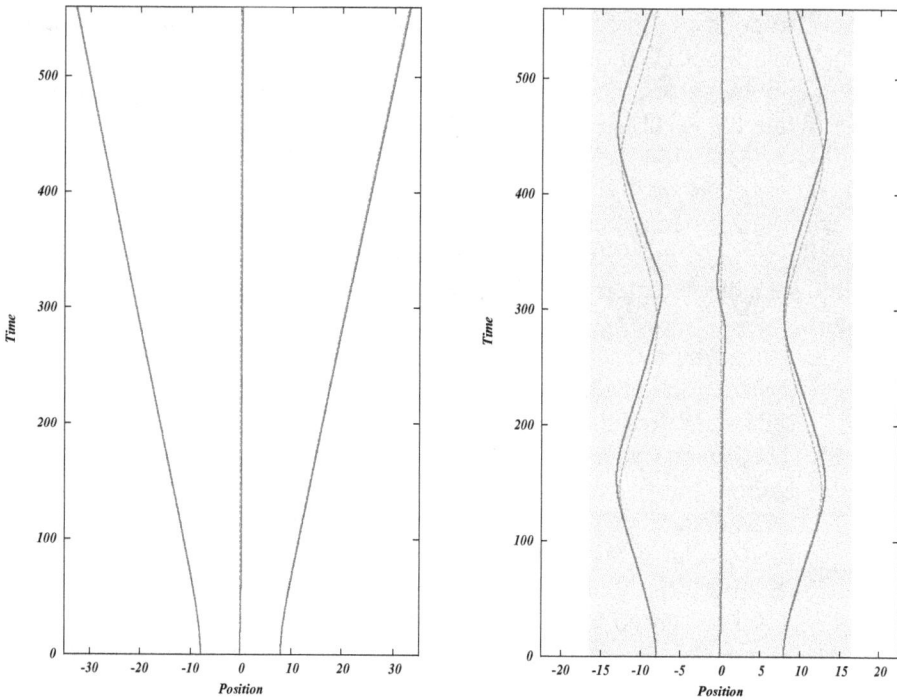

Figure 2.59. AFR: free potential behavior corresponding to real parts of eigenvalues of the Lax pair $\Re\zeta_1 = -0.0116$, $\Re\zeta_2 = 0$, $\Re\zeta_3 = 0.0116$ (left panel); external potential well $V(x) = \sum_{s=0}^{32} c_s \, \text{sech}^2(x - x_s)$, $c_s = -10^{-1}$, $x_s = -16 + s$, $s = 0, ..., 32$. The shaded area denotes the external potential at a half level (right panel).

a regime takes place if the amplitudes are chosen as in (2.81) and the phases are as in (2.84*b*); see the left panel of figure 2.60;

BSR Bound state regime when all solitons move asymptotically with the same velocity. Such a regime takes place for amplitudes with $\Delta\nu > \nu_{cr}$ and the phases are as in (2.84*a*). Such an asymptotic regime is shown in the left panel of figure 2.61.

It is natural to analyze separately all three regimes and to see what would be the effect of the external wells/humps on them. In particular, one can determine for which positions and intensities of the external potentials the solitons will undergo from one asymptotic regime to another.

Remark 1. *The CTC and its perturbative version PCTC use the adiabatic approximation. If we assume that the distance between the solitons is $r_0 = 8$, then the adiabatic parameter $\epsilon \simeq 0.01$, so one can expect that the CTC model will hold true up to times of the order of $1/\epsilon \simeq 100$. In figure 2.60, we see an excellent match between the MM and CTC up to 300 dimensionless time units; after that the two models diverge.*

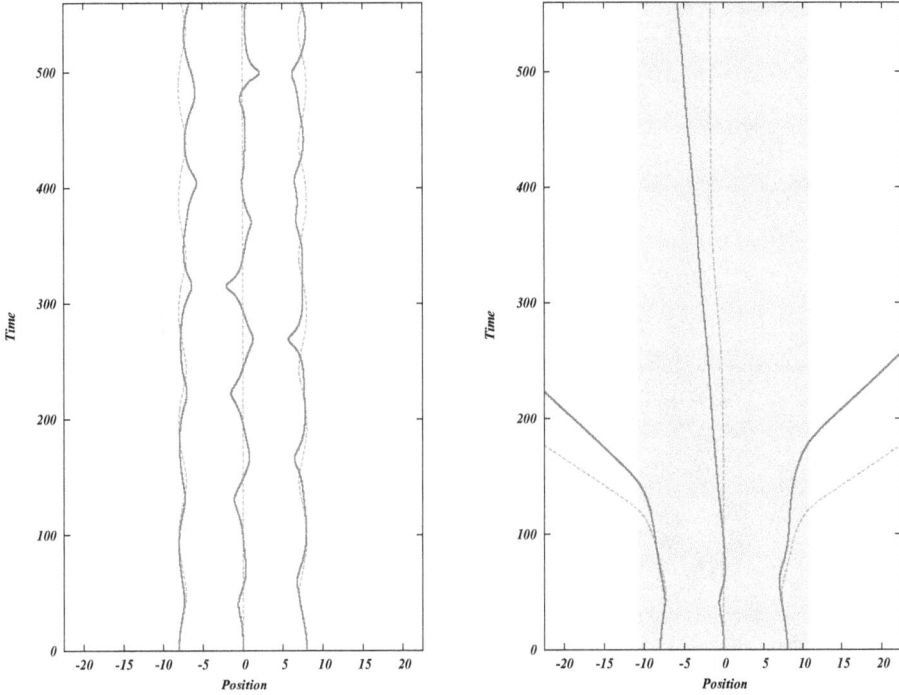

Figure 2.60. BSR: free potential behavior corresponding to real parts of eigenvalues of the Lax pair $\Re\zeta_1 = \Re\zeta_2 = \Re\zeta_3 = 0$ (left); external potential hump $V(x) = \sum_{s=0}^{12} c_s \operatorname{sech}^2(x - x_s)$, $c_s = 10^{-2}$, $x_s = -10 + sh$, $h = 5/3$, $s = 0, ..., 12$. The shaded area denotes the external potential at a half level (right).

Since the PCTC model is not integrable, we will solve it numerically to find the predicted solitons trajectories $\xi_k(t)$. Besides, we will solve numerically the Manakov model with the initial condition (2.56) and extract the trajectories of $\max(|u_1|^2 + |u_2|^2)$, where $\vec{u} \equiv (u_1, u_2)$.

In the right panel of figure 2.58, we plot samples of potential well with a width of 40 composed of 33 wells with depth $c_s = -0.1$ distributed uniformly between abscissas -16 and 16 and a distance between them of $h = 1$.

Evidently each Manakov soliton solution is parameterized by six parameters and four of them are the usual velocity, position, amplitude and phase. Two more parameters fix up the polarization vector. Keeping in mind the big parametric phenomenology of the solutions, we fix the velocities, positions and polarization vectors and vary the initial amplitudes and phases in order to ensure one or another asymptotic regime [38]. Even with only a three solitons configuration but with 13–33 potential wells/humps, we have a large variety of combinations.

Potential wells, especially when broad enough, attract the solitons and may be used to stabilize the solitons in a bound state. Potential humps repel the solitons; choosing their positions appropriately, one can either split a soliton bound state into free solitons or force free solitons into the bound state.

In what follows, we compare the PCTC models with the numerical solutions of the corresponding (perturbed) MS. In doing this, to have a better base for

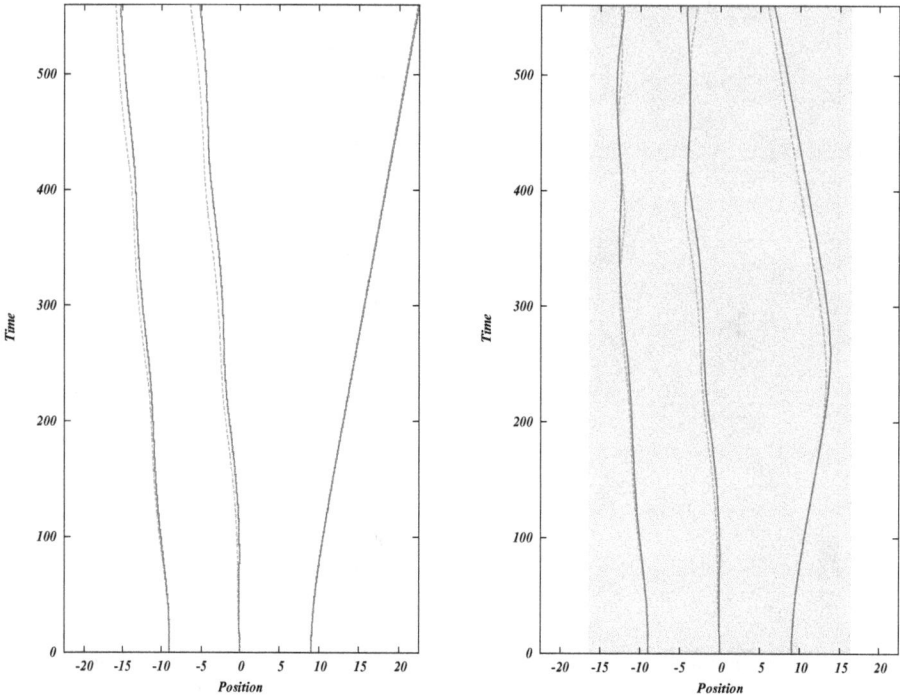

Figure 2.61. MAR: free potential behavior corresponding to real parts of eigenvalues of the Lax pair $\Re\zeta_1 = \Re\zeta_2 = -0.00321$, $\Re\zeta_3 = 0.00642$ (left); external potential well $V(x) = \sum_{s=0}^{32} c_s \operatorname{sech}^2(x - x_s)$, $c_s = -10^{-2}$, $x_s = -16 + sh$, $h = 1$, $s = 0, \ldots, 32$. The shaded area denotes the external potential at a half level (right).

comparison, we keep fixed some of the initial parameters of the soliton trains. The other parameters may vary from run to run; their particular values will be specified in the captions of the figures. To avoid any confusion, we mark the PCTC solutions by dashed lines, and the numerical solutions of the MS and the perturbed MS by solid lines. Also, we plot the centers of solitons and track their trajectories. Since the PCTC are derived in the framework of the adiabatic approximation, they are expected to be adequate only up to times of the order of ϵ^{-1}. So, if the distance between neighboring solitons is eight units, then $\epsilon \simeq 10^{-2}$ and one might expect that the PCTC would be valid up to $t \simeq 100$. Rather surprisingly, we find that the models work well until $t \simeq 1000$ or even longer. We also assume that $\xi_k < \xi_{k+1}$, $k = 1, 2$.

Below, we provide several examples that illustrate our points. More specifically, we set:

IC-1 The initial velocities $\mu_{k,0} = 0$, $k = 1, 2, 3$;

IC-2 The initial positions $\xi_{1,0} = -8$, $\xi_{2,0} = 0$, $\xi_{2,0} = 8$;

IC-3 Each of the initial polarization vectors $\vec{n}_{k,0}$ will be parameterized by its polarization angle $\theta_{k,0}$ and a phase $\gamma_{k,0}$ as follows:

$$\vec{n}_{k,0} = \begin{pmatrix} e^{i\gamma_{k,0}} \cos(\theta_{k,0}) \\ e^{-i\gamma_{k,0}} \sin(\theta_{k,0}) \end{pmatrix}. \tag{2.82}$$

Generically the scalar products $(\vec{n}^{\dagger}_{k+1}, \vec{n}_k)$ are complex-valued. For simplicity here, we assume that $\gamma_{k,0} = 0$ for $k = 1, 2, 3$ so that $(\vec{n}^{\dagger}_{k+1}, \vec{n}_k) = \cos(\theta_{k+1,0} - \theta_{k,0})$ and $\theta_{k,0} = (4 - k)\pi/10$. Thus, all scalar products just mentioned are equal to $\cos(\pi/10) \simeq 0.951$;

IC-4 The initial amplitudes

$$\nu_{1,0} = \nu_0 + \Delta\nu, \ \nu_{2,0} = \nu_0 = 0.5, \ \nu_{3,0} = \nu_0 - \Delta\nu; \tag{2.83}$$

IC-5 We use two types of initial phases configurations:

(a) $\delta_{1,0} = 0, \ \delta_{2,0} = \pi, \ \delta_{3,0} = 0, \quad \Delta\nu = 0.01.$

If $\Delta\nu < 0.02526$—asymptotically free behavior; $\tag{2.84a}$

(b) $\delta_{1,0} = \delta_{2,0} = \delta_{3,0} = 0, \quad \Delta\nu = 0.02.$

Bound state behavior for every $\Delta\nu > 0.$ $\tag{2.84b}$

Results and discussion

In the next figures, we show some examples of 3-soliton systems. In the figures, we plot the trajectories predicted by the PCTC (green dashed lines) with the Manakov model (red solid lines). In order to ensure the adiabaticity condition, we assume that initially the distance between the neighboring solitons $r_0 = 8$. The first example (figure 2.59) clearly demonstrates the role of the external well on the stability of the asymptotically free 3-soliton configuration. The potential (shaded strip) does not allow the lateral solitons to leave the well and they start to oscillate. So, the new regime is a bound state.

In figure 2.60, the potential free regime is a bound state (left panel). The influence of the potential hump of width 24 and amplitude $c_s = 10^{-2}$ (see the right panel) leads to fast violation of this regime and transitions to asymptotically free behavior of the lateral solitons.

Figure 2.61 demonstrates the influence of external potential on the third possible regime—the mixed asymptotic regime. In a potential free configuration, we have two bound stated solitons and one freely propagating. The adding of an external potential as superposed wells with amplitude (depth) $c_s = -10^{-2}$ leads to a bound state behavior of all the three solitons.

In these figures, the solutions obtained by VNLSE are plotted by red solid lines, while those obtained by CTC model by dashed green line. The comparison of the numerical predictions of both models is fully satisfactory.

Conclusions

We analyze the effects of the external potential wells and humps on the VNLSE soliton interactions using the PCTC model. An important future problem is applying the adiabatic approximation method also to the more general VNLSE model (2.2). For small values of cross-modulation α_2, it could also be consistently modeled by the corresponding PCTC [45].

The superposition of a large number of wells/humps obviously complexifies the motion of the soliton envelopes and can cause a transition from an asymptotically free and mixed asymptotic regime to a bound state regime and vice versa. In particular, the latter means that the external potentials can be used to control the soliton motion in a given direction and therefore to achieve a predicted motion of the optical pulse. A general feature of the conducted experiments is that the predictions of both models match very well for a very long-time evolution. This means that PCTC is a reliable model for predicting the evolution of the multisoliton solutions of the Manakov model in adiabatic approximation.

2.13 Manakov solitons and effects of external potential wells and humps

The superposition of a large number of wells/humps strongly influences the motion of the soliton envelopes and can cause a transition from asymptotically free and mixed asymptotic regime to a bound state regime and vice versa. Such external potentials are easier to implement in experiments and can be used to control the soliton motion in a given direction and to achieve a predicted motion of the optical pulse. A general feature of the conducted numerical experiments is that the predictions of both CTC and PCTC match very well with the Manakov model numerics for long-time evolution, often much longer than expected. This means that PCTC is a reliable *dynamical* model for predicting the evolution of the multisoliton solutions of the Manakov model in adiabatic approximation.

Below, we concentrate on wide but shallow *sech*-like potentials, equation (2.55) where $y_{s+1} - y_s = 1$ and the quantity A is large, so that initially the whole N-soliton train is in the potential well/hump (see figure 2.62). When $A \to \infty$ the wide well-type (hump-type) potentials can be presented by the integral

$$
\begin{aligned}
V_0(x, y_i, y_f) &= c_0 \int_{y_i}^{y_f} \frac{dy}{\cosh^2(2\nu_0(x - y))} \\
&= -\frac{c_0}{2\nu_0}[\tanh(2\nu_0(x - y_f)) - \tanh(2\nu_0(x - y_i))]
\end{aligned}
\tag{2.85}
$$

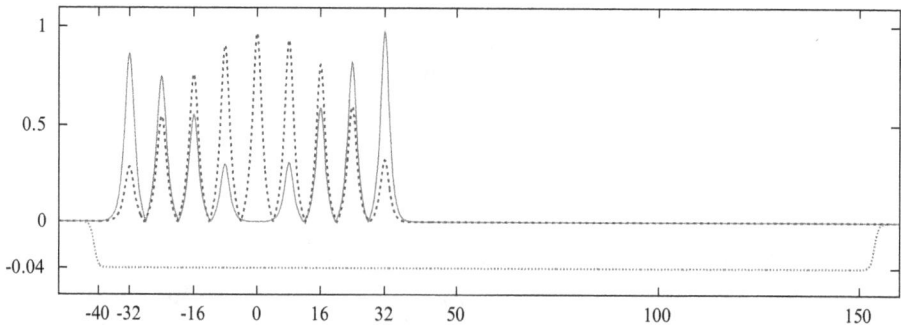

Figure 2.62. Graph of the external potential $V(x)$ from equation (2.85) (dashed pink line) and the initial configuration of soliton envelopes corresponding to figure 2.66 (left panel), and the modules of the first (in red) and second (in blue) components of $\vec{u}$.

where $c_0 < 0$, $a = y_f - y_i \gg 1$ is a large parameter. For $c_0 > 0$ ($c_0 < 0$), this will be a wide well-type (hump-type) potential.

We remind you that the basic idea of the adiabatic approximation is to derive a dynamical system for the soliton parameters which would describe their interaction. This idea was initiated by Karpman and Solov'ev [61] and modified by Anderson and Lisak [13]. Later, this idea was generalized to N-soliton interactions of scalar NLSE solitons [38, 39, 46] and then to the Manakov model; see [30, 33, 35, 40, 43].

The study is a natural extension of the results in [67] where combinations of potentials like in (2.55) were considered. We compare the numerical solutions of the perturbed Manakov model (2.54) with the predictions of the PCTC model. To this end, we solve the MS numerically by using an implicit scheme of Crank–Nicolson type in complex arithmetic, (2.5) and (2.6). The concept of the internal iterations is applied (see [23]) in order to ensure the implementation of the conservation laws on a difference level within the round-off error of the calculations [99, 100, 105]. The solutions of the relevant PCTC have been obtained by using the Runge–Kutta method of fifth order in complex arithmetic. Knowing the numeric solution $\vec{u}$ of the perturbed MS, we calculate the maxima of $(\vec{u}^{\dagger}, \vec{u})$, compare them with the numerical solutions for $\xi_k(t)$ of the PCTC, and plot the model trajectories predicted by both for each of the solitons.

The PCTC as a model for the soliton interaction of perturbed VNLSE

The CTC is an integrable dynamical model. This allows one to predict the asymptotic behavior of the solitons using the Lax representation $L = [B, L]$ of CTC (see equations (2.76)–(2.78)).

The PCTC taking into account the effects of the *sech*-like potentials to the best of our knowledge is not integrable and does not allow Lax representation; so we are applying numerical methods to solve it. Our main aim here is to find out potential configurations, which result in a transition from one asymptotic regime to another.

The effects of the sech-like and well-like potentials on CTC

The effects of the external potentials of the form (2.55) modifies the CTC to the following PCTC system like equation (2.73).

Choosing $V(x) = V_0(x)$ as in (2.85), we get:

$$
\begin{aligned}
P_0(\Delta) &= \int^{\Delta} d\Delta' P(\Delta') = \frac{\sinh(\Delta) - \Delta \cosh(\Delta)}{\sinh^3(\Delta)}, \\
R_0(\Delta) &= \int^{\Delta} d\Delta' R(\Delta') = = \frac{e^{-\Delta} \sinh^2(\Delta) + \Delta^2 \cosh(\Delta) - 2\Delta \sinh(\Delta)}{2 \sinh^3(\Delta)},
\end{aligned}
\tag{2.86}
$$

with $\Delta_{k,s} = 2\nu_0 \xi_k - y_s$ (figure 2.63). Then, M_k and D_k in (2.73) must be replaced by

$$
\begin{aligned}
M_{0,k} &= 2c_0\nu_k\big(P_0(z_k - y_f) - P_0(z_k - y_f)\big), \\
D_{0,k} &= \frac{c_0}{2\nu_0} \cdot \big(R_0(z_k - y_f) - R_0(z_k - y_f)\big).
\end{aligned}
\tag{2.87}
$$

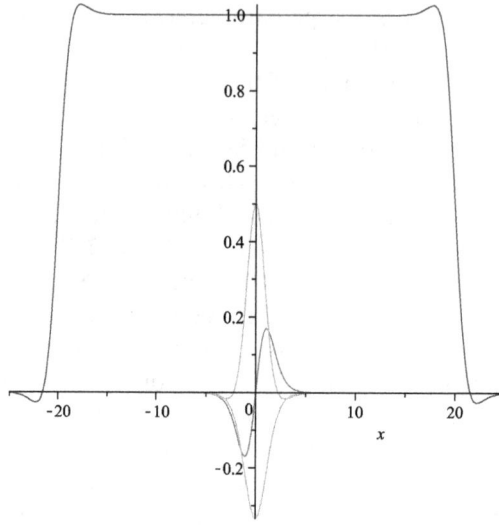

Figure 2.63. The functions $P(\Delta)$ (in red), $R(\Delta)$ (in green), $P_0(\Delta)$ (in orange) and $R_0(y_i) - R_0(y_f)$ for $y_f = -y_i = 20$ (in blue).

Comparison between the PCTC model and Manakov soliton interactions

In this section, we compare how well the PCTC model predicts the soliton interactions of the MS with the external potentials of kind (2.55).

It is natural to analyze separately all three regimes and see what would be the effect of the external wells/humps on them. In particular, one can determine for which positions and intensities of the external potentials the solitons will undergo from one asymptotic regime to another.

Remark 2. *The CTC and its perturbative version PCTC use the adiabatic approximation. If we assume that the distance between the solitons is $r_0 = 8$, then the adiabatic parameter $\epsilon_0 \simeq 0.01$, so one can expect that the CTC model will hold true up to times of the order of $1/\epsilon \simeq 100$. Rather surprisingly, quite often, we find that the models work well until $t \simeq 1000$ or even longer.*

We first demonstrate a plot of the wide well with the initial position of the nine Manakov solitons (figure 2.62).

Let us now specify the initial sets of soliton parameters, which we use in our tests in figure 2.64.

$$\xi_k = -45 + 9k, \quad \nu_k = 0.4625 + 0.0075(k-1), \quad \mu_k = 0,$$
$$\delta_k = k\pi, \quad \theta_k = \frac{(10-k)\pi}{10}, \quad k = 1, \ldots, 9. \tag{2.88}$$

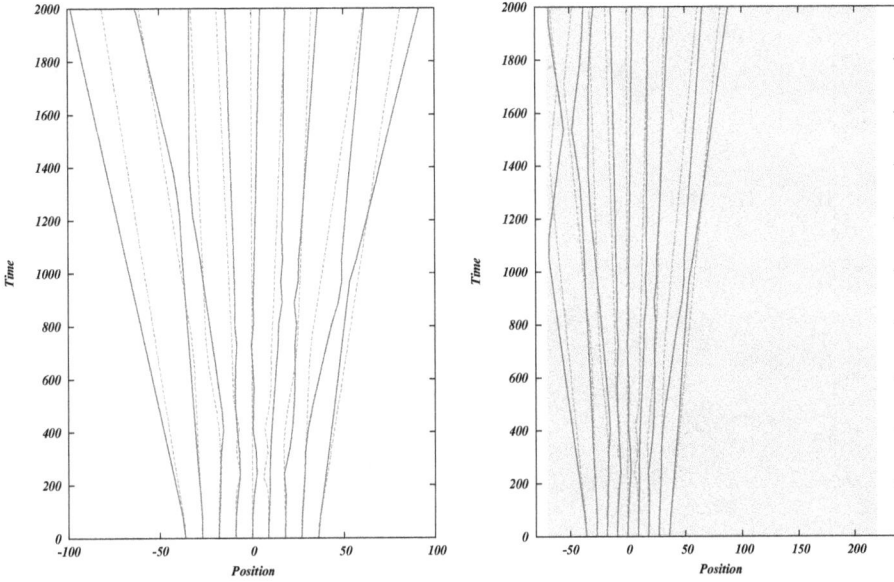

Figure 2.64. Trajectories of the soliton centers of potential-free nine-soliton asymptotic regime *red*— MS, *green*—CTC (left); trajectories of the soliton centers under external potential $V(x)$ (equation (2.85)) with $c_s = -0.004$, $y_i = -69.5$, $y_f = 218.5$, *red*—MS, *green*— CTC (right).

In addition, we assume that the components of the polarization vectors are real. The real parts of the eigenvalues of the relevant 9×9 Lax matrix are given by:

$$\kappa_1 = \kappa_2 = -0.005720, \qquad \kappa_3 = \kappa_4 = -0.001564, \qquad \kappa_5 = 0.0000,$$
$$\kappa_6 = \kappa_7 = 0.001564, \qquad \kappa_8 = \kappa_9 = 0.005720, \tag{2.89}$$

which means that there are four pairs of solitons moving with equal velocities and forming four bound states; the middle soliton in this frame of reference stays at rest; see the left panel of figure 2.64. Of course, the external potential destroys the asymptotic regime; see the right panel of figure 2.64.

For the next test, we use somewhat different initial conditions, namely now the distance between the neighboring solitons is 8:

$$\xi_k = -40 + 8k, \qquad \nu_k = 0.4625 + 0.0075(k-1), \qquad \mu_k = 0,$$
$$\delta_k = k\pi, \qquad \theta_k = \frac{(10-k)\pi}{10}, \qquad k = 1, \dots, 9. \tag{2.90}$$

The eigenvalues of the Lax matrix now are given by:

$$\kappa_1 = \kappa_2 = -0.011877, \qquad \kappa_3 = \kappa_4 = -0.006926, \qquad \kappa_5 = 0.0000,$$
$$\kappa_6 = \kappa_7 = 0.006926, \qquad \kappa_8 = \kappa_9 = 0.011877, \tag{2.91}$$

which means that we have again four pairs of 2-soliton bound states, now moving with larger velocities. The middle soliton is again at rest; see the left panel of figure 2.65.

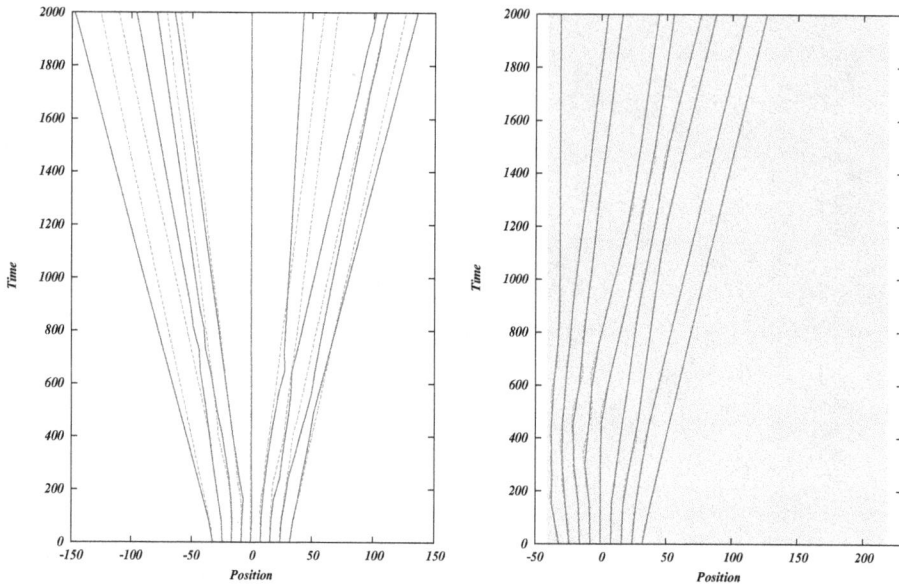

Figure 2.65. Trajectories of the soliton centers of potential free nine-soliton asymptotic regime *red*—MS, *green*—CTC (left); trajectories of the soliton centers under external potential $V(x)$ (equation (2.85)) with $c_s = -0.004$, $y_i = -40.5$, $y_f = 215.5$, *red*—MS, *green*—CTC (right).

The parameters in figures 2.66 are the same as in figure 2.65.

Evidently, each Manakov soliton solution is parameterized by six parameters and four of them are the usual velocity, position, amplitude and phase. Two more parameters fix up the polarization vector. Keeping in mind the big parametric phenomenology of the solutions, we fix the velocities, positions and polarization vectors and vary the initial amplitudes and phases in order to ensure one or another asymptotic regime [38]. Even with only three soliton configurations and wide potential wells/humps with $A = 13$ to $A = 33$, we have a large variety of combinations.

Potential wells, especially when broad enough attract the solitons and may be used to stabilize in a bound state. Potential humps repel the solitons; choosing their positions appropriately one can split a soliton bound state into free solitons.

In what follows, we compare the PCTC models with the numeric solutions of the corresponding (perturbed) Manakov model. We mark the PCTC solutions by dashed lines, and the numeric solutions of the Manakov model and the perturbed Manakov model by solid lines.

The comparison of the numerical predictions of both models in all asymptotic cases is very good.

Conclusions

The superposition of a large number of wells/humps strongly influences the motion of the soliton envelopes and can cause a transition from asymptotically free and

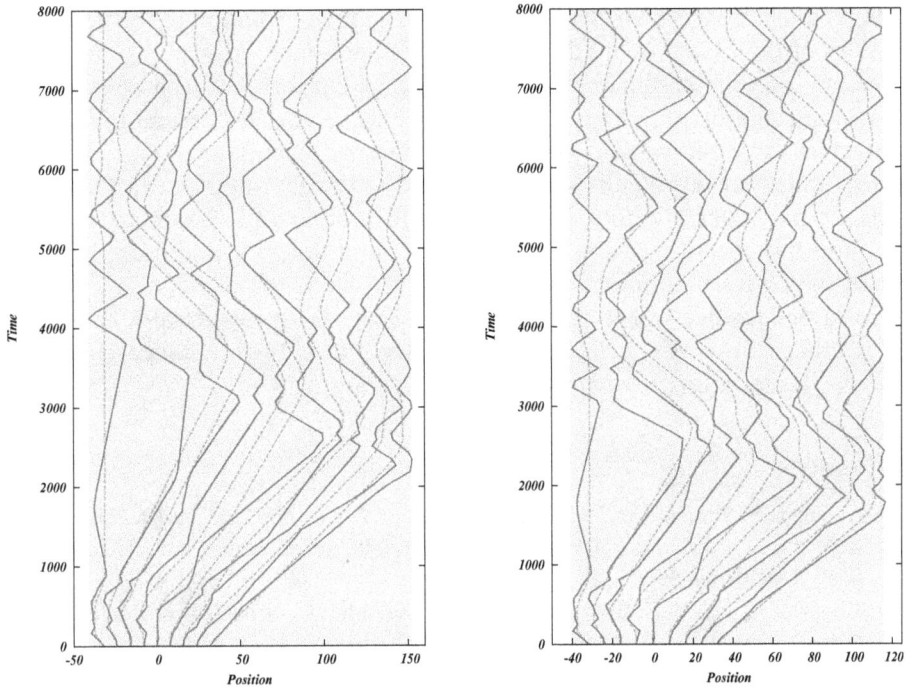

Figure 2.66. Trajectories of the soliton centers under external potential $V(x)$ (equation (2.85)) with $c_s = -0.004$, $y_i = -40.5$, $y_f = 151.5$ (left); $y_i = -40.5$, $y_f = 115.5$ (right), *red*—MS, *green*—CTC.

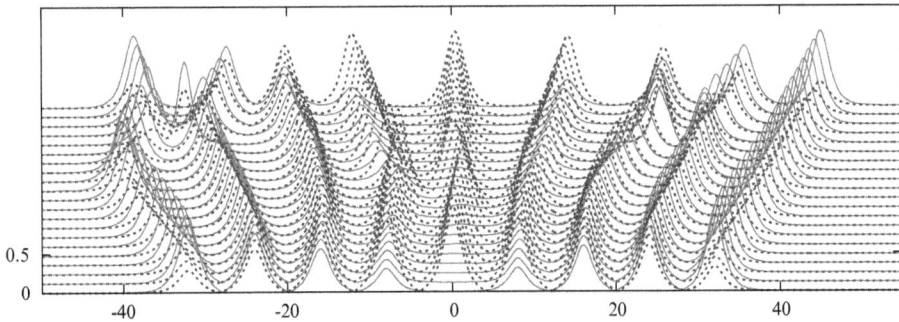

Figure 2.67. Dynamics of the solitons envelopes from figure 2.66 left up to time 320 (8000 time steps), *red*—module of the first component of $\vec{u}$, *blue*—module of the second component of $\vec{u}$. The prediction is given by numerical solution of perturbed MS. In this time interval, both models are in good agreement.

mixed asymptotic regimes to a bound state regime and vice versa. Such external potentials are easier to implement in experiments and can be used to control the soliton motion in a given direction and to achieve a predicted motion of the optical pulse. A general feature of the conducted numerical experiments is that the predictions of both CTC and PCTC match very well with the MM numerics for long-time evolution, often much longer than expected; see remark 2. This means that

PCTC is a reliable *dynamical* model for predicting the evolution of the multisoliton solutions of the Manakov model in adiabatic approximation.

We can consider infinite wells using $-c_1 V(x, y_i)$. I expect that such wells will affect the soliton chain by pushing it to one side; we will consider different initial soliton parameters and discuss additional effects due to the potentials.

Even rather weak potentials with $c_0 \simeq 10^{-4}$ can substantially affect the soliton trains. In particular, we found a special configuration of a 9-soliton train which asymptotically form four pairs of solitons in bound states symmetrically positioned around the central soliton.

We study the effects of the potential well, which is asymmetrical with respect to the soliton train. We can also study the effect of a 'staircase' potential $V_{sc}(x) = \sum_s c_s \tanh(2\nu_0(x - y_s))$, with arbitrary c_s and y_s.

References

[1] Abdulaev F, Darmanyan S and Khabibullaev P 1993 *Optical Solitons* (Berlin: Springer)

[2] Ablowitz M J and Ladik J F 1976 A nonlinear difference scheme and inverse scattering *Stud. Appl. Math.* **55** 213–29

[3] Ablowitz M J and Ladik J F 1976/77 On the solution of a class of nonlinear partial difference equations *Stud. Appl. Math.* **57** 1–12

[4] Ablowitz M J, Ohtab Y and Trubatch A D 1999 On discretizations of the vector nonlinear Schrödinger equation *Phys. Lett.* A **253** 287–304

[5] Ablowitz M J, Prinari B and Trubach A D 2003 On the IST for discrete nonlinear Schrödinger systems and polarization shift for discrete vector solitons *Nonlinear Physics— Theory and Experiment II Proc. Workshop 'Universite di Lecce—Consortium Einstein' Gallipoli, Italy 27 June–6 July 2002* (River Edge, NJ: World Scientific) pp 3–16

[6] Ablowitz M J, Prinari B and Trubach A D 2004 Discrete and continuous nonlinear Schrödinger systems *Lecture Notes Series* vol 302 (London: Cambridge University Press)

[7] Ablowitz M J and Segur H 1981 Solitons and the inverse scattering transform *SIAM Studies in Applied Mathematics* (Philadelphia, PA: SIAM)

[8] Agrawal G P 1997 *Fiber-optic Communication Systems* 2nd edn (New York: Wiley)

[9] Agrawal G P 1989 *Nonlinear Fiber Optics* (New York: Academic)

[10] Akhmediev N N and Ankiewicz A 1997 *Solitons, Nonlinear Pulses and Beams* (London: Chapman and Hall) p 39

[11] Akhmediev N N, Buryak V, Soto-Crespo J M and Anderson D R 1995 Phase-locked stationary soliton states in birefringent nonlinear optical fibers *J. Opt. Soc. Am.* B **12** 434–9

[12] Anderson D 1983 Variational approach to nonlinear pulse propagation in optical fibers *Phys. Rev.* A **27** 3135–45

[13] Anderson D and Lisak M 1983 Nonlinear asymmetric self-phase modulation and self-steepening of pulses in long optical waveguides *Phys. Rev.* A **27** 1393–8
Anderson D, Lisak M and Reichel T 1988 Approximate analytical approaches to nonlinear pulse propagation in optical fibers: a comparison *Phys. Rev.* A **38** 1618–20

[14] Baker S M, Elgin J N and Gibbons J 1999 Polarization dynamics of solitons in birefringent fibers *Phys. Rev.* E **62** 4325–32

[15] Blow K J, Doran N J and Wood D 1986 Polarization instabilities for solitons in birefringent fibers *Opt. Lett.* **12** 202–4

[16] Busch T and Anglin J R 2001 Dark-bright solitons in inhomogeneous Bose-Einstein condensates *Phys. Rev. Lett.* **87** 010401

[17] Chen Y 1998 Stability criterion of coupled soliton states *Phys. Rev.* E **57** 3542–50

[18] Chen Y and Haus H A 2000 Manakov solitons and polarization mode dispersion *Chaos* **10** 529–38

[19] Chen Y and Atai J 1995 Polarization instabilities in birefringent fibers: a comparison between continuous waves and solitons *Phys. Rev.* E **52** 3102–5

[20] Chow K W, Nakkeeran K and Malomed B 2003 Periodic waves in bimodal optical fibers *Opt. Commun.* **219** 251–9

[21] Christov C I 1994 Gaussian elimination with pivoting for multidiagonal systems *Internal Report* 4 University of Reading

[22] Christov I and Christov C I 2008 Physical dynamics of quasi-particles in nonlinear wave equations *Phys. Lett.* A **372** 841–8

[23] Christov C I, Dost S and Maugin G A 1994 Inelasticity of soliton collisions in system of coupled NLS equations *Phys. Scr.* **50** 449–54

[24] Collings B C, Cundiff S T, Akhmediev N N, Soto-Crespo J M, Bergman K and Knox W H 2000 Polarization-locked temporal vector solitons in a fiber laser: experiment *J. Opt. Soc. Am.* **B17** 354–65

[25] Delqué M, Fanjoux G and Sylvestre T 2007 Polarization dynamics of the fundamental vector soliton of isotropic Kerr media *Phys. Rev.* E **37** 016611

[26] Elgina J N, Enolskib V Z and Itsc A R 2007 Effective integration of the nonlinear vector Schrödinger equation *Physica* D **225** 127–52

[27] Ermakov V V and Kalitkin N N 1981 The optimal step and regularization for Newton's method *Sov. Comput. Phys. Math. Phys.* **21** 235

[28] Fornberg B 1996 *A Practical Guide to Pseudospectral Methods* (Cambridge: Cambridge University Press)

[29] Fornberg B and Whitham G B 1973 A numerical and theoretical study of certain nonlinear wave phenomena *Phil. Trans. R. Soc.* A **289** 373–404

[30] Gerdjikov V S 2011 Modeling soliton interactions of the perturbed vector nonlinear Schrödinger equation *Bulgarian J. Phys.* **38** 274–83

[31] Gerdjikov V S, Grahovski G G and Kostov N A 2005 On the multi-component NLS type equations on symmetric spaces and their reductions *Theor. Math. Phys.* **144** 1147–56

[32] Gerdjikov V S 1998 *N*-soliton interactions, the complex Toda chain and stability of NLS soliton trains *Proc. Int. Symp. on Electromagnetic Theory* vol 1 ed E Kriezis (Thessaloniki: Aristotle University of Thessaloniki) pp 307–9

[33] Gerdjikov V S, Baizakov B B and Salerno M 2005 Modelling adiabatic *N*-soliton interactions and perturbations *Theor. Math. Phys.* **144** 1138–46

[34] Gerdjikov V S, Baizakov B B, Salerno M and Kostov N A 2006 Adiabatic *N*-soliton interactions of Bose-Einstein condensates in external potentials *Phys. Rev.* E **73** 046606

[35] Gerdjikov V S, Doktorov E V and Matsuka N P 2007 *N*-soliton train and generalized complex Toda chain for Manakov system *Theor. Math. Phys.* **151** 762–73

[36] Gerdjikov V S, Evstatiev E G and Ivanov R I 1998 The complex Toda chains and the simple Lie algebras—solutions and large time asymptotics *J. Phys. A: Math. Gen.* **31** 8221–32

[37] Gerdjikov V S, Evstatiev E G and Ivanov R I 2000 The complex Toda chains and the simple Lie algebras—solutions and large time asymptotics—II *J. Phys. A: Math Gen.* **33** 975–1006

[38] Gerdjikov V S, Evstatiev E G, Kaup D J, Diankov G L and Uzunov I M 1998 Stability and quasi-equidistant propagation of NLS soliton trains *Phys. Lett.* A **241** 323–8

[39] Gerdjikov V S, Kaup D J, Uzunov I M and Evstatiev E G 1996 Asymptotic behavior of *N*-soliton trains of the nonlinear Schrödinger equation *Phys. Rev. Lett.* **77** 3943–6

[40] Gerdjikov V S, Kostov N A, Doktorov E V and Matsuka N P 2009 Generalized perturbed complex Toda chain for Manakov system and exact solutions of the Bose-Einstein mixtures *Math. Comput. Simul.* **80** 112–9

[41] Gerdjikov V S, Kostov N A and Valchev T I 2009 Bose-Einstein condensates with $F = 1$ and $F = 2$: reductions and soliton interactions of multi-component NLS models *Proc. SPIE* **7501** 75010W

[42] Gerdjikov V S and Todorov M D 2014 *N*-soliton interactions for the Manakov system. Effects of external potentials *Localized Excitations in Nonlinear Complex Systems, Nonlinear Systems and Complexity* ed J Carretero Gonzalez *et al* vol 7 (Bern: Springer) pp 147–69

[43] Gerdjikov V S and Todorov M D 2013 On the effects of sech-like potentials on Manakov solitons *AMiTaNS'13*AIP CP1561 ed M D Todorov (Melville, NY: AIP) pp 75–83

[44] Gerdjikov V S, Todorov M D and Kyuldjiev A V 2016 Asymptotic behavior of Manakov solitons: effects of potential wells and humps *Math. Comput. Simul.* **121** 166–78

[45] Gerdjikov V S, Todorov M D and Kyuldjiev A V 2017 Adiabatic interactions of Manakov solitons—effects of cross-modulation *Wave Motion* **71** 71–81

[46] Gerdjikov V S, Uzunov I M, Evstatiev E G and Diankov G L 1997 Nonlinear Schrödinger equation and *N*-soliton interactions: generalized Karpman-Soloviev approach and the complex Toda chain *Phys. Rev.* E **55** 6039–60

[47] Gerdjikov V S and Uzunov I M 2001 Adiabatic and non-adiabatic soliton interactions in nonlinear optics *Physica* D **152-153** 355–62

[48] Gerdjikov V S, Vilasi G and Yanovski A B 2008 Integrable Hamiltonian hierarchies. Spectral and geometric methods *Lecture Notes in Physics* vol 748 (Berlin: Springer)

[49] Gross E P 1961 Structure of a quantized vortex in boson systems *Il Nuovo Cimiento* **20** 454

[50] Griffin A, Nikuni T and Zaremba E 2009 *Bose-Condensed Gases at Finite Temperatures* (Cambridge: Cambridge University Press)

[51] Gross E P 1963 Hydrodynamics of a superfluid condensate *J. Math. Phys.* **4** 195–207

[52] Haus H A and Wong W S 1996 Solitons in optical communications *Rev. Modern Phys.* **68** 423–44

[53] Haelterman M and Sheppard A P 1994 The elliptically polarized fundamental vector solution of isotropic Kerr media *Phys. Lett.* **A194** 191–6

[54] Haelterman M and Sheppard A P 1994 The vector soliton associated with polarization modulational instability in the normal dispersion regime *Proc. 1994 Nonlinear Optics: Materials, Fundamentals, and Applications Waikoloa, HI 25–27 July 1994 IEEE* 10.1109/NLO.1994.470862

[55] Hempelmann U 1995 Polarization coupling and transverse interaction of spatial optical solitons in a slab waveguide *J. Opt. Soc. Am.* **B12** 77–86

[56] Ho T-L 1998 Spinor Bose condensates in optical traps *Phys. Rev. Lett.* **81** 742

[57] Ieda J, Miyakawa T and Wadati M 2004 Exact analysis of soliton dynamics in spinor Bose-Einstein condensates *Phys. Rev. Lett.* **93** 194102

[58] Ismail M S and Taha T R 2001 Numerical simulation of coupled nonlinear Schrödinger equation *Math. Comput. Simulat.* **56** 547–62

[59] Jensen S M The nonlinear coherent coupler *IEEE J. Quantum Electron.* **18** 1982 1580

[60] Kanna T and Lakshmanan M 2002 Effect of phase shift in shape changing collision of solitons in coupled nonlinear Schrödinger equations *Eur. Phys. J.* B **29** 249–54

[61] Karpman V I and Solov'ev V V 1981 A perturbational approach to the two-solition systems *Physica* D **3** 487–502

[62] Kevrekidis P G, Frantzeskakis D J and Carretero-Gonzalez R 2008 *Emergent Nonlinear Phenomena in Bose-Einstein Condensates: Theory and Experiment* vol 45 (Berlin: Springer)

[63] Kostov N A, Enol'skii V Z, Gerdjikov V S, Konotop V V and Salerno M 2004 On two-component Bose-Einstein condensates in periodic potential *Phys. Rev.* E **70** 056617

[64] Kostov N A and Gerdjikov V S 2007 On the soliton interactions in NLS equation with external potentials *Proc. SPIE* **6604** 66041S

[65] Kostov N A, Gerdjikov V S and Valchev T I 2007 Exact solutions for equations of Bose-Fermi mixtures in one-dimensional optical lattice *SIGMA* **3** 07114

[66] Kovachev L M 1992 Influence of cross phase modulation and four-photon parametric mixing on the relative motion of optical pulses *Opt. Quantum Electron.* **24** 1992

[67] Kyuldjiev A V, Gerdjikov V S and Todorov M D 2014 Asymptotic behavior of Manakov solitons: effects of shallow and wide potential wells and humps *Mathematics in Industry* ed A Slavova (Cambridge: Cambridge Scholar Publishing) pp 410–26

[68] Lake B M, Yuen H C, Rungaldier H and Ferguson W E 1977 Nonlinear deep-water waves; theory and experiment. Part 2. Evolution of a continuous wave train *J. Fluid Mech.* **83** 49–74

[69] Lakoba T I, Kaup D J and Malomed B A 1997 Solitons in nonlinear fiber couplers with two orthogonal polarizations *Phys. Rev.* E **55** 6107–20

[70] Lakoba T I and Kaup D J 1997 Perturbation theory for the Manakov soliton and its applications to pulse propagation in randomly birefringent fibers *Phys. Rev.* E **56** 6147–65

[71] Malomed B A 2002 Variational methods in nonlinear fiber optics and related fields *Prog. Opt.* **43** 171–93

[72] Manakov S V 1974 On the theory of two-dimensional stationary self-focusing of electro-magnetic waves *Sov. Phys. JETP* **38** 248–53

[73] Marcuse D, Menyuk C and Wai P 1997 Application of the Manakov-PMD equation to studies of signal propagation in optical fibers with randomly varying birefringence *J. Lightwave Technol.* **15** 1735–46

[74] Menyuk C R 1989 Pulse propagation in an elliptically birefringent Kerr medium *IEEE J. Quantum Electr.* **25** 2674–82

[75] Menyuk C 2004 Interaction of nonlinearity and polarization mode dispersion *J. Opt. Fibers Commun. Rep.* **1** 305–11

[76] Menyuk C and Marks B 2006 Interaction of polarization mode dispersion and nonlinearity in optical fiber transmission systems *J. Lightwave Technol.* **24** 2806–26

[77] Modugno M, Dalfovo F, Fort C, Maddaloni P and Minardi F 2000 Dynamics of two colliding Bose-Einstein condensates in an elongated magnetostatic trap *Phys. Rev.* A **62** 063607

[78] Moser J 1975 Dynamical systems, theory and applications *Lecture Notes in Physics* vol 38 (Berlin: Springer) p 467

[79] Newell A C 1985 *Solitons in Mathematics and Physics* (Philadelphia, PA: SIAM)

[80] Novikov S P, Manakov S V, Pitaevski L P and Zakharov V E 1984 *Theory of Solitons, The Inverse Scattering Method* (New York: Consultant Bureau)

[81] Ohmi T and Machida K 1998 Bose-Einstein condensation with internal degrees of freedom in alkali atom gases *J. Phys. Soc. Jpn.* **67** 1822

[82] Orstrovskaya E A, Akhmediev N N, Stegeman G I, Kang J U and Aichison J S 1997 Mixed-mode spatial solitons in semiconductor waveguides *J. Opt. Soc. Am.* B **14** 880

[83] Perez-Garcia V M, Michinel H, Cirac J I, Lewenstein M and Zoller P 1997 Dynamics of Bose-Einstein condensates: variational solutions of the Gross-Pitaevskii equations *Phys. Rev.* A **56** 1424–32

[84] Pitaevskii L P 1961 Vortex lines in an imperfect bose gas *Sov. Phys. JETP* **13** 451–4

[85] Pitaevskii L P and Stringari S 2003 *Bose-Einstein Condensation* (Oxford: Oxford University Press)

[86] Radhakrishnan R, Lakshmanan M and Hietarinta J 1997 Inelastic collision and switching of coupled bright solitons in optical fibers *Phys. Rev.* E **56** 2213–6

[87] Rand D, Glesk I, Brès C -S, Nolan D A, Chen X, Koh J, Fleischer J W, Steiglitz K and Prucnal P R 2007 Observation of temporal vector soliton propagation and collision in birefringent fiber *Phys. Rev. Lett.* **98** 053902

[88] Romagnoli M, Trillo S and Wabnitz S 1992 Soliton switching in nonlinear couplers *Opt. Quantum Electron.* **24** S1237–67

[89] Shchesnovich V S and Doktorov E V 1997 Perturbation theory for solitons of the Manakov system *Phys. Rev.* E **55** 7626

[90] Shrira V I and Geogjaev V V 2010 What makes the Peregrine soliton so special as a prototype of freak waves? *J. Eng. Math.* **67** 11–22

[91] Snyder A W, Chen Y, Rowland D and Mitchell D J 1990 Unification of nonlinear-optical fiber devices *Opt. Lett.* **15** 171–3

[92] Sonnier W J and Christov C I 2005 Strong coupling of Schrödinger equations: conservative scheme approach *Math. Comput. Simul.* **69** 514–25

[93] Sonnier W J and Christov C I 2009 Repelling soliton collisions in coupled Schrödinger equations with negative cross modulation *Discrete Cont. Dyn. Syst. Suppl.* **2009** 708–18

[94] Sonnier W J 2011 Dynamics of repelling soliton collisions in coupled Schrödinger equations *Wave Motion* **48** 805–13

[95] Sophocleus C and Parker D F 1994 Pulse collisions and polarisation for optical fibres *Opt. Comm.* **112** 214–24

[96] Taha T R and Ablowitz M J 1984 Analytical and numerical aspects of certain nonlinear evolution equations. I. Analytical *J. Comput. Phys.* **55** 192–202

[97] Taha T R and Ablowitz M J 1984 Analytical and numerical aspects of certain nonlinear evolution equations. II. Numerical Schrödinger equation *J. Comput. Phys.* **55** 203–30

[98] Toda M 1989 *Theory of Nonlinear Lattices* (Berlin: Springer)

[99] Todorov M D and Christov C I 2007 Conservative scheme in complex arithmetic for coupled nonlinear Schrödinger equations *Discrete Cont. Dyn. Syst. Suppl.* **2007** 982–92

[100] Todorov M D and Christov C I 2009 Impact of the large cross-modulation parameter on the collision dynamics of quasi-particles governed by vector NLSE *Math. Comput. Simulat.* **80** 46–55

[101] Todorov M D and Christov C I 2009 Collision dynamics of circularly polarized solitons in nonintegrable coupled nonlinear Schrödinger system *Discrete Cont. Dyn. Syst. Suppl.* **2009** 780–9

[102] Todorov M D and Christov C I 2008 On the solution of the system of ODEs governing the polarized stationary solutions of CNLSE *3rd Annual Session of BGSIAM, 22–23 December 2008* Hosted by Institute of Math. and Informatics, BAS (Sofia: Demetra Ltd.) pp 83–6

[103] Todorov M D and Christov C I 2011 Collision dynamics of polarized solitons in linearly CNLSE *Int Workshop on Complex Structures, Integrability and Vector Fields* AIP CP1340 (Melville, NY: AIP) pp 144–53

[104] Todorov M D 2011 Polarization dynamics during takeover collisions of solitons in systems of coupled nonlinear Schrödinger equations *Discrete Cont. Dyn. Syst. Suppl.* **2011** 1385–94

[105] Todorov M D and Christov C I 2012 Collision dynamics of elliptically polarized solitons in coupled nonlinear Schrödinger equations *Math. Comput. Simul.* **82** 1321–32

[106] Todorov M D 2016 The effect of the elliptic polarization on the quasi-particle dynamics of linearly coupled systems of nonlinear Schrödinger equations *Math. Comput. Simul.* **127** 273–86

[107] Uchiyama M, Ieda J and Wadati M 2007 Multicomponent bright solitons in $F = 2$ spinor Bose-Einstein cndensates *J. Phys. Soc. Jpn.* **76** 74005

[108] Ueda M and Koashi M 2002 Theory of spin-2 Bose-Einstein condensates: spin correlations, magnetic response, and excitation spectra *Phys. Rev.* A **65** 063602

[109] Uzunov I M, Gerdjikov V S, Gölles M and Lederer F 1996 On the description of N-soliton interaction in optical fibers *Opt. Commun.* **125** 237–42

[110] Uzunov I M, Muschall R, Gölles M, Kivshar Y S, Malomed B A and Lederer F 1995 Pulse switching in nonlinear fiber directional couplers *Phys. Rev.* E **51** 2527–37

[111] Winful H G 1985 Self-induced polarization changes in birefringent optical fibers *Appl. Phys. Lett.* **47** 213–5

[112] Yang J 1997 Classification of the solitary waves in coupled nonlinear Schrödinger equations *Phys. D* **108** 92–112

[113] Yang J 2002 Suppression of Manakov soliton interference in optical fibers *Phys. Rev.* E **65** 036606

[114] Zakharov V E and Shabat A B 1972 Exact theory of two-dimensional self focusing and one-dimensional self-modulation of waves in nonlinear media *Sov. Phys. JETP* **34** 62–9

[115] Zakharov V E and Shabat A B 1973 Interaction between solitons in a stable medium *Sov. Phys. JETP* **37** 823

Chapter 3

Ultrashort optical pulses. Envelope dispersive equations

We investigate the wave dynamics of ultrashort optical pulses via numerical implementation and computer visualition of the obtained results. The experimental observation and measurement of ultrashort pulses in waveguides is a hard job and this is the reason and stimulus to create mathematical models for computer simulations, as well as reliable algorithms for treating the governing equations.

The spatiotemporal evolution of a femtosecond pulse ($1\text{fs} = 10^{-15}$ s) through a nonlinear media is composed by condensed argon, as well as condensed helium and neon. The governing equation is (3+1)D nonlinear Schrödinger equation ((3+1)D NLSE). We split the linear part of the operator consecutively by physical factors and coordinates [22] and apply the internal iterations [7] for the nonlinear terms. The equation describes the diffraction, the group velocity dispersion (GVD), and the Kerr nonlinearity. The numerical simulations allow us to 'observe' the dynamics of self-focusing and splitting of the pulse. Under certain conditions, the intensity, transverse width and spatiotemporal shape of the pulse remain stable looking, like a soliton [25]. A selfcompression of the laser pulse in media with a positive dispersion is obtained [11]. When the laser pulse propagating in condensed helium and neon is a multi-millijoule femtosecond, one observes a drastic increase of the intensity of the pulse center [24].

3.1 On a method for solving of multidimensional equations of mathematical physics

We consider linear multidimensional evolutionary equations or the linear part of nonlinear ones [22]. The complex structure and the presence of terms with different physical senses requires the coordinate splitting to be preceded by splitting by physical factors (processes). In contrast to the coordinate splitting, this kind of splitting can be exact in some nodes and in the intervals it can be controlled. The method in question is developed in the 1980s by Marchuk [13] and until now it was applied very successfully for solving various problems in ecology, air and water pollution, diffusion, etc. Here, we show that this method is relevant for study of propagating ultrashort localized pulses in nonlinear waveguides as well. We aim to demonstrate that the splitting by physical factors is applicable and can be efficient for solving of multidimensional evolutionary problems in optics. These equations possess very rich phenomenology, containing nonlinear terms and they do not admit analytic solutions that are nonintegrable in the general case. Without losing generality, we pay attention to an initial-value problem of (3+1)D equations of the Schrödinger kind. We split the linear part of the differential operator and the initial conditions to two consequent Cauchy problems and by using a spectral analysis and techniques we prove that the original and the resulting splitting problems are equivalent in the ends of a given interval. For constant coefficients, the above assertion holds in the whole interval. In the opposite case for a small enough step, the approximation error varies in an acceptable range. The nonlinear terms are linearized by using so-called internal iteration.

The method is applied successfully for numerical solving of a (3+1)D Schrödinger equation with a sign-variable group velocity, as well for (3+1)D amplitude equations (of envelope), when the propagation regimes of the light pulses may be ultra-short (femtosecond). The obtained results are reliable and give good predictions for the material quantities and dynamics of the light pulses.

(3+1)D Schrödinger equation

Let us consider the Cauchy problem

$$i\frac{\partial u}{\partial z} + \frac{1}{2k}\left(\frac{\partial^2 u}{\partial x^2} + \frac{\partial^2 u}{\partial y^2}\right) - \frac{\beta}{2}\frac{\partial^2 u}{\partial \tau^2} = 0,$$

$$u(x, y, \tau, z = 0) = g(x, y, \tau), \quad u|_\Gamma = 0$$

(3.1)

where Γ is the boundary of 3D spatial region, g is a given function, $\tau = t - \frac{z}{v_g}$ is the local time in a moving frame with group velocity of the pulse v_g, k is the wave number, β is the dispersion, and z is the evolutionary coordinate of the pulse. Equation (3.1) describes two different physical processes: the first one is the diffraction of the pulse represented by the equation

$$i\frac{\partial u}{\partial z} + \frac{1}{2k}\left(\frac{\partial^2 u}{\partial x^2} + \frac{\partial^2 u}{\partial y^2}\right) = 0$$

$$u(x, y, \tau, z = 0) = g(x, y, \tau), u|_\Gamma = 0.$$

The second process relates to the dispersion and is governed by the equation

$$i\frac{\partial u}{\partial z} - \frac{\beta}{2}\frac{\partial^2 u}{\partial \tau^2} = 0$$

$$u(x, y, \tau, z = 0) = g(x, y, \tau), u|_\Gamma = 0.$$

When the initial function and the sought solution approach fast zero uniformly in all arguments, there exist Fourier integrals

$$g = \int_{-\infty}^{\infty} \int_{-\infty}^{\infty} \int_{-\infty}^{\infty} \hat{g}(m, n, p, z) \exp[i(mx + ny + p\tau)]\, dm\, dn\, dp \qquad (3.2)$$

$$u = \int_{-\infty}^{\infty} \int_{-\infty}^{\infty} \int_{-\infty}^{\infty} \hat{u}(m, n, p, z) \exp[i(mx + ny + p\tau)]\, dm\, dn\, dp \qquad (3.3)$$

where $\hat{g}$ and $\hat{u}$ are the Fourier images of the functions g and u.

On the other hand, the triple Fourier transform of equation (3.1) with respect to x, y, and τ gives a Cauchy problem of the ordinary differential equation

$$i\frac{\partial \hat{u}}{\partial z} - \frac{1}{2k}(m^2 \hat{u} + n^2 \hat{u}) + \frac{\beta}{2}p^2 \hat{u} = 0 \quad \hat{u}(z = 0) = \hat{g},$$

which admits the exact solution

$$\hat{u} = \hat{g} \exp\left[i\left(-\frac{1}{2k}m^2 - \frac{1}{2k}n^2 + \frac{\beta}{2}p^2\right)\right]. \qquad (3.4)$$

Then, the solution of the original problem, equation (3.1), has the following integral form

$$u = \int_{-\infty}^{\infty} \int_{-\infty}^{\infty} \int_{-\infty}^{\infty} \hat{g}(m, n, p)$$
$$\times \exp\left\{i\left[mx + ny + p\tau - \left(\frac{m^2}{2k} + \frac{n^2}{2k} - \frac{\beta}{2}p^2\right)z\right]\right\}\, dm\, dn\, dp. \qquad (3.5)$$

Fourier analysis. Splitting by physical processes

Let us introduce an interval $0 \leqslant z \leqslant \xi$ and consider the following initial-value problems

$$i\frac{\partial u_1}{\partial z} + \frac{1}{2k}\left(\frac{\partial^2 u_1}{\partial x^2} + \frac{\partial^2 u_1}{\partial y^2}\right) = 0 \qquad (3.6)$$
$$u_1(x, y, \tau, z = 0) = g(x, y, \tau)$$

and

$$i\frac{\partial u}{\partial z} - \frac{\beta}{2}\frac{\partial^2 u}{\partial \tau^2} = 0 \qquad (3.7)$$
$$u(x, y, \tau, z = \xi) = u_1(x, y, \tau, \xi).$$

3-3

If function g decades on infinity, then the Fourier-images of the above problems exist. They admit exact solutions

$$\hat{u}_1(z) = \hat{g} \exp\left[-\frac{i}{2k}(m^2 + n^2)z\right]$$

$$\text{and} \quad \hat{u}(z) = \hat{u}_1(\xi) \exp\left(i\frac{\beta_2}{2}p^2 z\right).$$

(3.8)

After obvious algebraic manipulations, we obtain

$$u = \int_{-\infty}^{\infty} \int_{-\infty}^{\infty} \int_{-\infty}^{\infty} \hat{g}(m, n, p)$$

$$\times \exp\left\{i\left[mx + ny + p\tau - \left(\frac{m^2}{2k} + \frac{n^2}{2k} - \frac{\beta}{2}p^2\right)\xi\right]\right\} dm\, dn\, dp.$$

(3.9)

For $z = \xi$, the solutions (3.6) and (3.10) are equivalent for any ξ while for $0 \leqslant z \leqslant \xi$, the solution of the split problem is approximate. The latter means that in a discrete set of nodes z_j, $j = 0, 1, 2, \ldots$ one has to consequently solve the initial-value problems

$$i\frac{\partial u_1}{\partial z} + \frac{1}{2k}\left(\frac{\partial^2 u_1}{\partial x^2} + \frac{\partial^2 u_1}{\partial y^2}\right) = 0$$

$$u_1(x, y, \tau, z = z_j) = u^j$$

(3.10)

and

$$i\frac{\partial u}{\partial z} - \frac{\beta}{2}\frac{\partial^2 u}{\partial \tau^2} = 0$$

$$u(x, y, \tau, z = z_j) = u_1^{j+1}.$$

(3.11)

This is the gist of the method of splitting by physical processes applied for multidimensional problems.

By using Fourier analysis, we showed that the original problem (3.1) is equivalent to consequently solve initial-value problems (3.11) and (3.12) at a given set of nodes z_j, $j = 0, 1, 2, \ldots$. In the concrete case, (3.1) is (2+1)D and needs one more splitting—this time coordinate and after that it can be solved numerically by the alternative directions implicit (ADI) method, e.g. by the Peaceman–Rachford scheme or by the scheme of stabilizing correction [29]. In opposition, equation (3.12) is (1+1)D and can be solved direct by the Crank–Nicolson scheme in complex arithmetic. The resulting three equations possess sparse matrices and can be easily treated by the Thomas algorithm and its complex-valued modifications [6, 23].

Splitting by physical processes is applied for more a complicated equation (3.1) with cubic nonlinearity and an initial condition—function $g(x, y, \tau) \equiv \exp(-\frac{x^2}{2} - \frac{y^2}{2} - \frac{\tau^2}{2})$. To also get an implicit scheme about the nonlinear part, we use so-called internal iteration, which is convergent for small enough steps in the z-direction [7, 21].

(3+1)D envelope equation

Let us consider one more Schrödinger-like equation—the envelope equation, which is a (3+1)D dispersion equation with cubic, quintic and rational nonlinearities. The linear part of the equation is

$$i\frac{\partial u}{\partial z} + \frac{1}{2k}\Delta u - ia_1\frac{\partial}{\partial \tau}\Delta u - a_2\frac{\partial^2 u}{\partial \tau^2} - ia_3\frac{\partial^3 u}{\partial \tau^3} = 0, \tag{3.12}$$

where $\Delta = \dfrac{\partial^2}{\partial x^2} + \dfrac{\partial^2}{\partial y^2}$, a_1, a_2, a_3 are dimensionless real constants. For simplicity, we will not comment on their physical sense in this paper. The splitting by physical processes leads to the following two initial-value problems

$$i\frac{\partial u}{\partial z} + \frac{1}{2k}\Delta u - i\frac{a_1}{2}\frac{\partial}{\partial \tau}\Delta u = 0$$
$$u(x, y, \tau, z = 0) = g, \quad u|_\Gamma = 0 \tag{3.13}$$

and

$$i\frac{\partial u}{\partial z} - i\frac{a_1}{2}\frac{\partial}{\partial \tau}\Delta u - a_2\frac{\partial^2 u}{\partial \tau^2} - ia_3\frac{\partial^3 u}{\partial \tau^3} = 0$$
$$u(x, y, \tau, z = 0) = g, \quad u|_\Gamma = 0. \tag{3.14}$$

The Fourier analysis results in consecutive solving in a discrete set of nodes $z_j, j = 0, 1, 2, \ldots$ of initial-value problems

$$i\frac{\partial u_1}{\partial z} + \frac{1}{2k}\Delta u_1 - ia_1\frac{\partial}{\partial \tau}\Delta u_1 = 0$$
$$u_1(x, y, \tau, z = z_j) = u_j \tag{3.15}$$

and

$$i\frac{\partial u}{\partial z} - a_2\frac{\partial^2 u}{\partial \tau^2} - ia_3\frac{\partial^3 u}{\partial \tau^3} = 0$$
$$u(x, y, \tau, z = z_j) = u_1^{j+1}. \tag{3.16}$$

Equation (3.16) requires a coordinate splitting, while equation (3.17) can be solved directly by a complex-valued Crank–Nicolson scheme. Due to the higher dispersion, the band matrix is four-diagonal. In order to use the Thomas algorithm again, modifications of the mixed derivative in both problems should be put on the old z-stage. In this way, the scheme of stabilizing correction for equation (3.16) yields

$$i\frac{\tilde{u}-u^j}{\Delta z/2} = -\frac{1}{2k}(\Lambda_{xx}\tilde{u} + \Lambda_{yy}u^j) + ia_1(\Lambda_{\tau xx} + \Lambda_{\tau yy})u^j,$$
$$i\frac{u^{j+1} - \tilde{u}}{\Delta z/2} = -\Lambda_{yy}(u^{j+1} - u^j). \tag{3.17}$$

Here, $(\tilde{\cdot})$ stands for the values of the grid function on the z-half-stage, $(\cdot)^j$—old z-stage $(\cdot)^{j+1}$—new z-stage, $\Lambda_{xx} \approx \dfrac{\partial^2}{\partial x^2}$, $\Lambda_{yy} \approx \dfrac{\partial^2}{\partial y^2}$, $\Lambda_{\tau xx} \approx \dfrac{\partial^3}{\partial \tau \partial x^2}$, $\Lambda_{\tau yy} \approx \dfrac{\partial^3}{\partial \tau \partial y^2}$. The first equation is implicit in the x-direction, while the second one—in the y-direction. The fully fledged (3+1)D nonlinear envelope equation is quite more complicated compared to the (3+1)D nonlinear Schrödinger equation

$$
\begin{aligned}
&i\frac{\partial u}{\partial z} + \frac{1}{2k}\Delta u - ia_1\frac{\partial}{\partial \tau}\Delta u - a_2\frac{\partial^2 u}{\partial \tau^2} - ia_3\frac{\partial^3 u}{\partial \tau^3} \\
&+ b_1\,|u|^2 u + ib_2\frac{\partial}{\partial \tau}(|u|^2 u) - b_3\,|u|^4 u - i\frac{\partial}{\partial \tau}(|u|^4 u) \\
&+ i\rho u + ib_6\frac{\partial}{\partial \tau}(\rho u) + ib_7\frac{u}{|u|^2} = 0.
\end{aligned}
\tag{3.18}
$$

The last three terms describe the ionization processes, the cubic and quintic non-linearities—the effects of self-steepening, the spatial self-focusing (self-defocusing), b_1, …, b_7 are real constant coefficients, ρ is the electron concentration, which is a dynamical quantity. The latter requires us to solve equation (3.19) together with the kinetic equation

$$
\frac{\partial \rho}{\partial \tau} = t_1(u)\rho + t_2, \quad t_1\text{—functional coefficient}, \quad t_2\text{—real constant}.
\tag{3.19}
$$

The nonlinearities can be linearized again by an internal iteration convergent for small enough z-steps. Equation (3.20) being 1D for given values of set function u is treated by the Crank–Nicolson scheme. An important advantage of the linear parts of the above splitting schemes is their absolute stability and full approximation of the differential operators.

The above considered equations are femtosecond (1fs $= 10^{-15}$ s). In order to emphasize the significance of these equations, we stress that they govern and are a tool for the study of the spatio-temporal dynamics of the molecular and atomic processes.

Conclusions

We implement the above splitting in finite differences for the (3+1)D nonlinear Schrödinger equation and (3+1) nonlinear envelope equation considering real regimes of pulse propagation. Here, we do not intend to describe the physical results because they are considered thoroughly in the next sections [11, 26, 27]. We aim only to motivate the splitting by physical processes combined with the internal iteration as a reliable method for solving multidimensional wave and amplitude equations governing ultrashort regimes, when exact solutions are impossible and the experimental observations and measurements are difficult and expensive.

3.2 Dynamics of high-intensity ultrashort light pulses at some basic propagation regimes

General behavior of high-intensity ultrashort light pulses

The propagation of high-intensity ultrashort light pulses (HULPs) is a complex phenomenon due to the common action of a large number of physical processes. They can be put into linear and nonlinear processes:

The main linear processes are dispersion and diffraction.

The nonlinear processes are due to neutrals and plasma (figure 3.1).

General characteristics of optical processes

The linear processes develop gradually with the pulse parameters τ and w.

The nonlinear processes in neutrals develop rapidly (depending on the order of nonlinearity $\chi^{(3)}, \chi^{(5)}, \chi^{(7)}, \ldots$) with the pulse parameters: I (intensity).

The nonlinear processes in plasma develop almost abruptly (depending on the order of nonlinearity) with the pulse parameters: I (intensity).

The nonlinearity that rules the ionization is much higher than the leading nonlinearities in neutrals, e.g. $\chi^{(21)}$, for the case of 11-photon ionization, as in the case of our studies.

For short and intense pulses, like the HULPs, the linear and nonlinear processes are very strong simultaneously.

The common action of all processes dramatically intensifies the pulse rearrangement that, in turn, affects back the shape and the parameters of the pulse. Thus, the prediction of the pulse behavior becomes a challenging problem that is not yet fully investigated and solved.

The common action of all processes results in a number of propagation regimes. Some of the most important will be presented in this work.

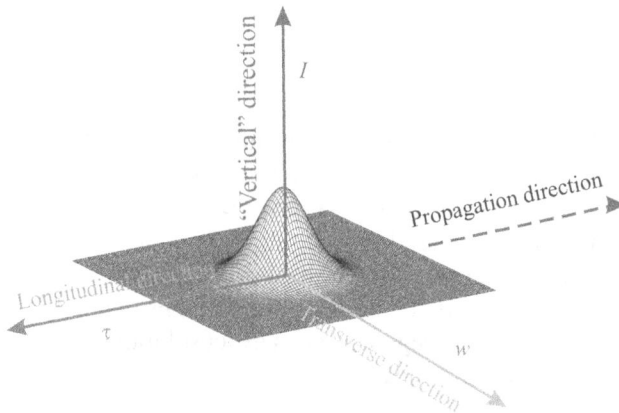

Figure 3.1. The intensiveness of above processes varies with the parameters of the pulse: pulse duration τ, transversal width w, and peak intensity I.

The aim of the present work

To summarize the spatiotemporal dynamics of HULPs at some basic propagation regimes that can be reached by the pulses generated from the most widely used high-power ultrafast lasers.

The following main cases have been considered:
1. (3+1)D nonlinear Schrödinger equation (NLSE)—ionization free regime.
2. (3+1)D nonlinear envelope equation (NEE)—ionization free regime.
3. (3+1)D nonlinear envelope equation (NEE)—ionization regime.

The laser pulses fall in the millijoule energy range having from around one hundred to a few tens of femtoseconds ($1\text{fs} = 10^{-15}$ s) time duration and a few tens of gigawatts peak power.

Realistic physical conditions have been only used throughout our studies!

3.3 (3+1)D nonlinear Schrödinger equation

The propagation of high-intensity femtosecond laser pulses in a bulk medium is a complex phenomenon accompanied by a strong spatiotemporal and spectral rearrangement of the pulse. The rigorous description of the pulse propagation is based on the (3+1)D nonlinear Schrödinger equation (NLSE), in which diffraction, group velocity dispersion of second order, and Kerr nonlinearity of third order form a *basic set of optical processes* of lowest order [5, 8, 17–19]. Additional terms in the NLSE are required to describe some particular features or more extreme conditions in the pulse propagation, such as, e.g. self-steepening, space–time focusing, non-instantaneous nonlinearity, higher order dispersion, etc [3, 16, 30]. Higher orders of nonlinearity and ionization must also be considered for pulses creating significant ionization [14]. A simultaneous collapse of the pulse in space and time and formation of a spatiotemporal soliton are predicted at negative GVD [19], while, in the more common case of positive GVD, the theoretical and experimental studies predict time broadening and splitting of the pulse [5, 8, 16–18, 30]. Self-compression (SC), before splitting, of high-intensity femtosecond laser pulses in positive GVD has been found in various types of media—pure atomic, molecular gases and fused silica [12]. The generation of compressed (almost five times in magnitude) pulses is accompanied by a strong increase of the peak intensity, an improvement of the spatiotemporal pulse shape, and a stable propagation of the compressed pulse [12].

The (3+1)D NLSE in moving frame for the complex field amplitude $\tilde{E}(r, z, \tau)$, where r is a polar coordinate [11, 24]

$$\frac{\partial \tilde{E}}{\partial z} - \frac{i}{2k}\nabla_\perp^2 \tilde{E} + \frac{i\beta_2}{2}\frac{\partial^2 \tilde{E}}{\partial \tau^2} - \frac{ikn_2}{n_0}|\tilde{E}|^2\tilde{E} = 0 \qquad (3.20)$$

is used as a propagation equation. The last three terms in equation (3.21) form the basic physical processes: diffraction, dispersion, cubic nonlinearity, where $\nabla_\perp^2$ is the radial Laplacian. The initial pulse is a linearly polarized chirp-free Gaussian in space and in time of axial symmetry,

$$\tilde{E}(r, z = 0, \tau) = E_0 \exp\left(-\frac{r^2}{2r_0^2} - \frac{\tau^2}{2\tau_0^2}\right).$$

The lack of substantial loses means that the condition for conservation of the pulse energy holds, i.e.

$$W(z) = \int_{S_\infty} \int_{-\infty}^{\infty} |\tilde{E}(r, z, \tau)|^2 ds d\tau = W_0 = \text{const.}$$

Let $u = \tilde{E}/E_0$. Supposing that function $u \to 0$ uniformly with respect to transverse coordinate ϕ when $\tau \to \pm\infty$ and $r \to \infty$, and the solution is regular in the origin, i.e. $\partial u/\partial r = 0$ when $r = 0$ after straightforward calculations, we derive two conservation laws: for the mass

$$\frac{d}{dz} \int_V r |u|^2 dV = 0, \quad V = V(r, \phi, \tau)$$

and

$$\frac{d}{dz} \int_V r\left(-\frac{1}{2}\left|\frac{\partial u}{\partial r}\right|^2 + \beta\left|\frac{\partial u}{\partial \tau}\right|^2 + \gamma |u|^4\right) dV = 0$$

for the energy. Then the implementation of (3.11) and (3.12) in polar coordinates results in the following dimensionless partial differential equations

$$i\frac{\partial u}{\partial z} = -\frac{1}{4r}\frac{\partial}{\partial r}\left(r\frac{\partial u}{\partial r}\right) - \frac{\gamma}{4}|u|^2 u$$

$$i\frac{\partial u}{\partial z} = \frac{\beta}{2}\frac{\partial^2 u}{\partial \tau^2} - \frac{\gamma}{4}|u|^2 u,$$

(3.21)

where $\beta = L_{DF}/L_{DS}$, $\gamma = L_{DF}/L_{NL}$, L_{DF} is the diffraction length, L_{DS} is the dispersion length, and L_{NL} is the nonlinear length. The finite-difference approximation of each of the differential equations is by using and combining a Crank–Nicolson implicit scheme for the linear part and an internal iteration for the nonlinear

$$i\frac{u_{k,l}^{n+\frac{1}{2}} - u_{k,l}^n}{h_z} = -\frac{1}{8r_l h_r^2}\left[\left(r_{l-\frac{1}{2}}u_{k,l-1}^{n+\frac{1}{2}} - 2r_l u_{k,l}^{n+\frac{1}{2}} + r_{l+\frac{1}{2}}u_{k,l+1}^{n+\frac{1}{2}}\right)\right.$$

$$\left. + \left(r_{l-\frac{1}{2}}u_{k,l-1}^n - 2r_l u_{k,l}^n + r_{l+\frac{1}{2}}u_{k,l+1}^n\right)\right] - \frac{\gamma}{4}\left|u_{k,l}^{n+\frac{1}{2}, s}\right|^2 u_{k,l}^{n+\frac{1}{2}, s}$$

$$i\frac{u_{k,l}^{n+1} - u_{k,l}^{n+\frac{1}{2}}}{h_z} = \frac{\beta}{4h_\tau^2}\left[\left(u_{k-1,l}^{n+1} - 2u_{k,l}^{n+1} + u_{k+1,l}^{n+1}\right)\right.$$

$$\left. + \left(u_{k-1,l}^{n+\frac{1}{2}} - 2u_{k,l}^{n+\frac{1}{2}} + u_{k+1,l}^{n+\frac{1}{2}}\right)\right] - \frac{\gamma}{4}\left|u_{k,l}^{n+1, s}\right|^2 u_{k,l}^{n+1, s}.$$

Here, the set function $u_{k,l}^n \equiv u(r_l, \tau_k, z_n)$,

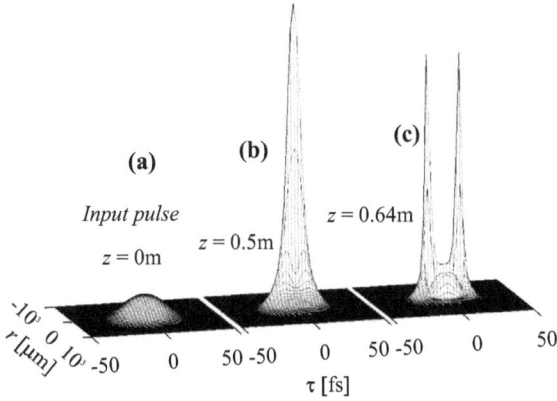

Figure 3.2. Pulse compression of high-energy femtosecond laser pulses in the positive GVD medium. Generation of optical 'tsunami.'

$$r_l = (l - 1)h_r, \quad h_r = \frac{r_{\text{inf}}}{N - 1}, \quad l = 1, \dots, N;$$

$$\tau_k = -\tau_{\text{inf}} + (k - 1)h_\tau, \quad h_\tau = \frac{2\tau_{\text{inf}}}{M - 1}, \quad k = 1, \dots, M.$$

Results and conclusions

Within this model, the pulse propagation has been studied in a number of atomic gases, Ar, Ne and He, at high pressure. Similar behavior has been obtained in all three cases, and illustrated in figure 3.2. Due to the self-focusing, the pulse intensity (a) grows up and the pulse undergoes self-compression in time (b), followed by splitting of the pulse (c). The pulse intensity grows up very rapidly around the point of maximal compression. Such a behavior has been called *optical tsunami*.

This study outlines the *minimal set* of processes at which self-compression can be observed, namely: dispersion, diffraction, and cubic nonlinearity.

- Compression of the pulse in time has been obtained at a minimal set of physical processes that includes diffraction, dispersion and cubic nonlinearity.
- Splitting of the pulse is observed immediately after the maximal time compression.
- The pulse does not show the signature of stabilization within the (3+1)D nonlinear Schrödinger equation. Generation of compressed pulses of rapid increase peak intensity around given space points and soliton-like intensity profile.

3.4 (3+1)D nonlinear envelope equation (NEE)

It is known that the model of (3+1)D NLSE is not able to describe a number of effects and properties of the power femtosecond laser pulses [2–4, 14]. The involvement of additional realistic factors can significantly increase the prediction for the real behavior of the pulses and hence the (3+1)D NLSE model is not enough and needs a further extension. The presence of nonlinearities of higher order and ionization can stabilize the temporal soliton-like pulse even for a positive dispersion [20]. This is the reason to use the (3+1)D nonlinear envelope equation (NEE) as a tool in our next investigations. The latter involves the description of single-cycled ultrashort light pulses in a strong ionization regime, which can influence the nonlinearity of the waveguide.

One way to achieve an ionization-free regime is simply to cancel the ionization terms in the self-consistent system, equations (3.23), (3.24), and (3.25) (see below). We will follow, however, another approach: the ionization-free regime will be achieved by a proper choice of HULP and medium parameters keeping the model unchanged. The latter will give information about the real physical parameters leading to such a regime. Also, it allows smooth switching from the ionization to ionization-free regime and vice versa simply by varying the pulse and medium parameters. Here, we solve *the full model, including ionization*, while the ionization-free regime (in fact, non-zero but negligible ionization) is achieved by a proper choice of the pulse and material parameters. The advantages of such an approach are twofold:

1. It allows smooth switching from the ionization-free to ionization regime and vice versa.
2. It allows us to find the real conditions at which both regimes can be achieved.

We follow the basic physical concept developed in [2]. The electromagnetic (optical) pulse is described in terms of a carrier-envelope concept: electric field of the pulse $E(t) = A(t)\exp[-i\omega_0 t + i\varphi(t)]$ + c.c. [3] with $A(t)$—amplitude, ω_0—carrier frequency, $\varphi(t)$—carrier-envelope phase. The above concept holds for pulses as short as a single-cycle period 2.5 fs at 800 nm.

The basic governing (3+1)D NEE was derived in 1997 [3] and legalized in 2007 [2]. It reads

$$\frac{\partial A}{\partial z} = \frac{i}{2k_0}\hat{T}^{-1}\nabla_\perp^2 A + i\hat{D}A$$

$$+ i\frac{\omega_0}{c}n_2\hat{T}\left[(1-x)|A|^2 + x\int_{-\infty}^t h(t-t')|A(t')|^2 dt'\right]A \qquad (3.22)$$

$$- i\frac{\omega_0}{c}n_4\hat{T}|A|^4 A - i\frac{k_0}{2n_0^2\rho_c}\hat{T}^{-1}\rho A - \frac{\sigma}{2}\rho A - \frac{\beta_{\text{MPI}}(A)}{2}A.$$

The first term in the right-hand side is diffraction and space–time focusing; the second one is the dispersion with the dispersion operator $\hat{D} = \sum_{n\geqslant 2}^{\infty}\frac{i^n}{n!}k^{(n)}\partial_t^n$; the third term combines the third order of nonlinearity (instantaneous and non-instantaneous) and self-steepening; the fourth one is fifth order of nonlinearity and self-steepening; the last three terms describe the ionization, with the sixth one being the ionization modification of the refractive index, the seventh one being the collision ionization, and the eighth one being the multi-photon/tunneling ionization. The third and fourth terms contain the self-steepening operator $\hat{T} = 1 + \frac{i}{\omega_0}\partial_t$, while the first and fifth ones contain the space–time focusing operator $\hat{T}^{-1} \approx 1 - \frac{i}{\omega_0}\partial_t$.

The above equation is coupled with the kinetic equation [2] about the electron number density ρ

$$\frac{\partial \rho}{\partial t} = W(I)(\rho_n - \rho) + \frac{\sigma(\omega_0)}{I_p}|A|^2\rho - f(\rho) \tag{3.23}$$

with three terms in the right hand side being the photoionization, the collision ionization by inverse bremsstrahlung, and the electron recombination. The ionization rate is in the multiphoton ionization regime, i.e. $W_{\mathrm{MPI}} = \sigma_K I^K$ with σ_K as the multiphoton ionization cross section, and K is the number of photons required to ionize the atom from its ground state. For Ar atoms (used here) at 800nm, $K = 11$ (figure 3.3). The ionization modification of the group velocity dispersion (GVD) is as that predicted in [10]. The total GVD is $k''_{\mathrm{total}} = k''_{\mathrm{tneutrals}} + k''_{\mathrm{ionization}}$ and

$$k''_{\mathrm{ionization}} = -\frac{e^2\lambda^3\rho}{2\pi^2 m_e c^4\left(1 - \frac{e^2\lambda^2\rho}{\pi m_e c^2}\right)^{\frac{3}{2}}}. \tag{3.24}$$

Thus, we arrive at a *minimal sufficient model*. In the minimal sufficient model, each strong physical process corresponds to at least one other strong physical process acting in the opposite direction.

In the case in question, the highest dispersion order we consider is equal to three and the coordinate system is again polar (like in (3+1)D NLSE). Then, the original equation (3.23) attains to the following dimensionless form

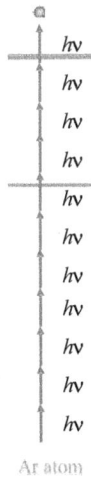

Figure 3.3. 11-photon absorption is required to ionize the Ar atom from the ground state.

$$\frac{\partial u}{\partial z} = \frac{i}{r}\frac{\partial}{\partial r}\left(r\frac{\partial u}{\partial r}\right) + a_1\frac{\partial}{\partial \tau}\left[\frac{1}{r}\left(r\frac{\partial u}{\partial r}\right)\right] - ia_2\frac{\partial^2 u}{\partial \tau^2}$$

$$+ a_3\frac{\partial^3 u}{\partial \tau^3} + ib_1\,|u|^2 u - b_2\frac{\partial}{\partial \tau}(|u|^2 u) - ib_3\,|u|^4 u + b_4\frac{\partial}{\partial \tau}(|u|^4 u) \tag{3.25}$$

$$- a_4\rho u - a_1\rho\frac{\partial u}{\partial \tau} - a_1 u\frac{\partial \rho}{\partial \tau} - b_5\frac{u}{|u|^2}.$$

Similarly, equation (3.24) attains to the dimensionless form

$$\frac{\partial \rho}{\partial \tau} = c_1\rho + c_2. \tag{3.26}$$

The coefficients

$$a_1 = \frac{1}{\omega_0\tau_0}, \quad a_2 = \frac{2L_{\mathrm{DF}}k^{(2)}}{\tau_0^2}, \quad a_3 = \frac{2L_{\mathrm{DF}}k^{(3)}}{3\tau_0^3}, \quad a_4 = i + \frac{\sigma n_0^2\rho_c}{k_0},$$

$$b_1 = \frac{4L_{\mathrm{DF}}\omega_0 n_2|A_0|^2}{c}, \quad b_2 = \frac{4L_{\mathrm{DF}}n_2|A_0|^2}{c\tau_0}, \quad b_3 = \frac{4L_{\mathrm{DF}}\omega_0 n_4|A_0|^4}{c},$$

$$b_4 = \frac{4L_{\mathrm{DF}}n_4|A_0|^4}{c\tau_0}, \quad b_5 = \frac{2L_{\mathrm{DF}}U_i W(I)}{|A_0|^2}\left(\rho_0 - \frac{n_0^2\rho_c}{2L_{\mathrm{DF}}k_0}\rho\right),$$

$$c_1 = \left(\frac{\sigma}{U_i}\,|A_0|^2|u|^2 - W(I)\right)\tau_0, \quad c_2 = \frac{2L_{\mathrm{DF}}k_0\tau_0}{n_0^2\rho_c}W(I)\rho_0$$

are material constants, τ_0, ω_0 are the duration and the transverse spatial width of the initial pulse, L_{DF} is the diffraction length, and $u = \frac{A}{A_0}$, $\tau = \frac{t}{\tau_0}$.

We split equation (3.26) by physical processes and get the following three PDEs

$$\frac{\partial u}{\partial z} = -ia_2\frac{\partial^2 u}{\partial \tau^2} + a_3\frac{\partial^3 u}{\partial \tau^3} + a_1\frac{\partial}{\partial \tau}\left[\frac{1}{r}\left(r\frac{\partial u}{\partial r}\right)\right]$$

$$+ ib_1\,|u|^2 u - b_2\frac{\partial}{\partial \tau}(|u|^2 u) - ib_3\,|u|^4 u + b_4\frac{\partial}{\partial \tau}(|u|^4 u)$$

$$\frac{\partial u}{\partial z} = \frac{i}{r}\frac{\partial}{\partial r}\left(r\frac{\partial u}{\partial r}\right) \tag{3.27}$$

$$\frac{\partial u}{\partial z} = -a_1\left(\rho\frac{\partial u}{\partial \tau} + u\frac{\partial \rho}{\partial \tau}\right) - a_4\rho u - b_5\frac{u}{|u|^2}.$$

The initial and the boundary conditions, and the grid are the same like in the case of (3+1)D NLSE and we will not repeat them. The finite-difference approximation of each of the differential equations and an implementation of the Crank–Nicolson implicit scheme for the linear parts of each of them is reduced to consecutive inversion of four-diagonal band matrices [6] and internal iterations for the nonlinear parts [7]. For small enough z-steps, the iteration converges within a few iterations.

Ionization-free regime

The main material parameters of the ionization-free regime used in the numerical calculations are summarized below: the initial pulse duration $\tau = 150$ fs, initial pulse energy $w = 0.5$ mJ, initial pulse diameter $D = 500$ μm, and the medium is gaseous argon at 3 atm. To verify the validity of the ionization-free regime at the above condition, the electron number density created by the pulse propagation is presented in figure 3.4 [28]. As can be seen, less than four electrons per cubic centimeter are predicted throughout the propagation distance. Such a low number density of the free electrons cannot produce any substantial effect on the pulse propagation. To analyze the stability of the pulse, we will characterize the pulse by two criteria—quantitative and qualitative ones. By the first one, we will characterize quantitatively the pulse by the following parameters: pulse duration τ (full width at half maximum (FWHM) of the temporal profile of the pulse), pulse diameter $2w$ (FWHM of the transversal profile of the pulse), and peak intensity I. We will also characterize qualitatively the pulse by means of stability of the temporal profile (the main problem for the pulse stability) as well as the entire spatiotemporal profile of the pulse.

The evolution of the pulse parameters, pulse diameter ($2w$), peak intensity I, and pulse duration τ is shown in figures 3.5, 3.6, and 3.7, respectively. The evolution of

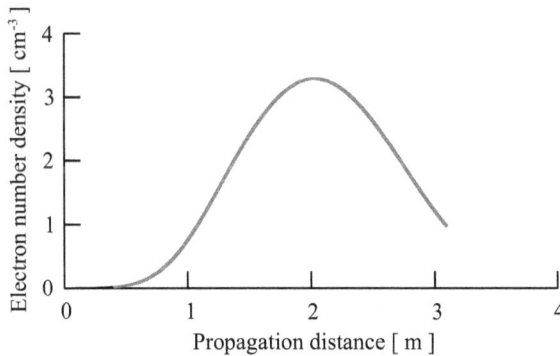

Figure 3.4. Electron number density versus propagation distance.

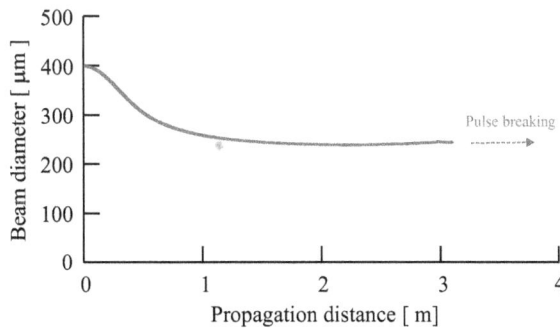

Figure 3.5. Beam diameter versus propagation distance. 1. Self-focusing and formation of light filament.

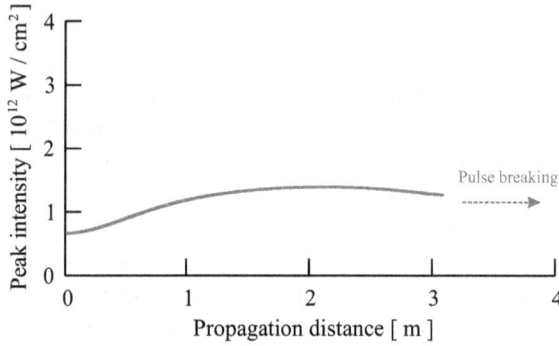

Figure 3.6. Peak intensity versus propagation distance. 2. Peak-intensity increasing.

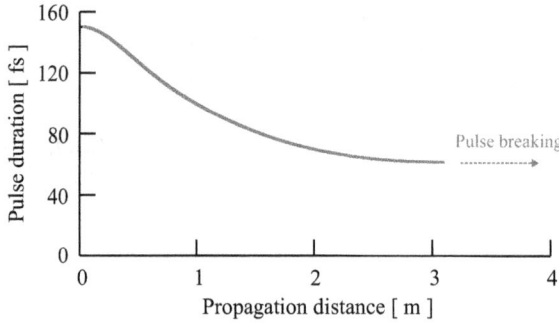

Figure 3.7. Pulse duration versus propagation distance. 3. Self-compression of the pulse in time.

the pulse in these figures results from the following set of processes. The pulse rearrangement begins with self-focusing due to the cubic nonlinearity. It leads to a radial flow of energy towards the center of the pulse, which shrinks the pulse in the transversal direction, figure 3.5. As a result, the peak intensity increases, figure 3.6, and, due to the low dispersion of the gaseous medium used here, it leads to effective shortening of the pulse in the time domain by a factor of two, figure 3.7. The shrinkage of the pulse in the transversal direction and increasing of the peak intensity increase the defocusing due to diffraction and fifth order of nonlinearity. At a given stage, a dynamic balance between the self-focusing and defocusing takes place, thus leading to stabilization of the pulse diameter along a distance L_w, figure 3.5. As no more effective flow of energy to the center of the pulse occurs once the radial balance is achieved, the peak intensity is also stabilized along a distance L_I, figure 3.6. Such stabilization is additionally assisted by the low dispersion of the medium, which lead to insubstantial expansion of the pulse along a distance in the order of L_I, being much shorter than the dispersion length L_{DS}. Due to the last reason, the pulse duration also remains stable along a distance L_τ, figure 3.7. The lengths of stability of the pulse are: $L_w = 0.91$ m, $L_I = 0.61$ m, $L_\tau = 0.36$ m. The lengths of stability are shown in figures 3.5–3.7 with bars, whose size and position are in scale. The pulse is most stable with respect to its width $2w$ and less stable with respect to its duration τ. The lengths of stability of the pulse do not, in general, coincide in space. In particular, the position of L_τ is substantially shifted in space with respect to L_W and L_I, figures 3.5–3.7. The characteristic lengths, calculated on the basis of the local parameters of the pulse in the area of stabilization are: $L_{DF} = 0.2$ m, $L_{DS} = 60$ m, $L_{NL} = 0.30$ m (figures 3.8–3.10). By comparison between the above lengths, one may conclude that all pulse parameters satisfy the condition of stability with respect to the diffraction length, i.e. L_w, L_I, $L_\tau > L_{DF}$, and the pulse can be considered as a *spatial soliton*. The same also

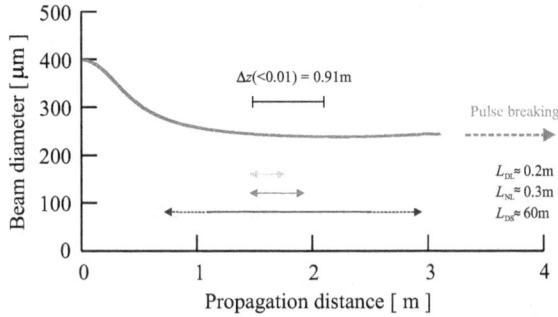

Figure 3.8. Beam diameter versus propagation distance. The distance at which the variation of given parameter is less than one percent with the pulse propagation is shown with black bar. The dispersion, diffraction and nonlinear characteristic lengths are shown in color bars.

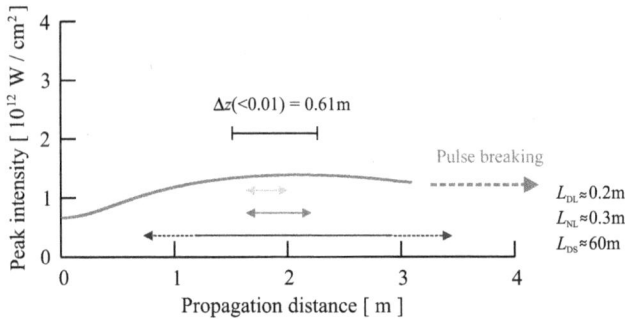

Figure 3.9. Peak intensity versus propagation distance.

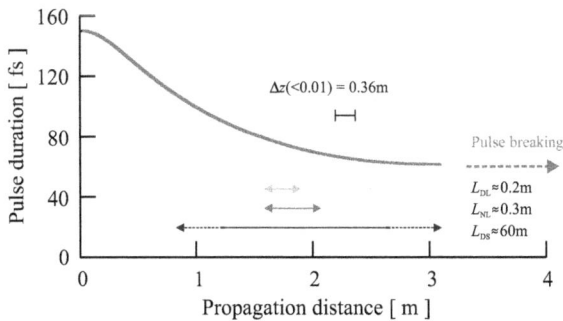

Figure 3.10. Pulse duration versus propagation distance.

holds with respect to the nonlinear length, L_w, L_I, $L_\tau > L_{NL}$. On the other hand, the pulse does not obey the condition of stability of the temporal soliton as the following relation holds: L_w, L_I, $L_\tau < L_{DS}$. Slightly more than 3 m from the beginning, the pulse breaking/splitting takes place. It is shown by a broken line in figures 3.5–3.7. The pulse splitting (from an experimental point of view) is considered as the pulse deterioration and the evolution of the pulse has not been followed at longer propagation distances. Apart from the parameters of the pulse, the HULP becomes stable (within given limits) with respect to its temporal profile, figure 3.11(a), and the entire spatiotemporal shape of the pulse, figure 3.11(b). As can be seen, the temporal profile and the spatiotemporal profile of the pulse do not practically change from 2.1 m to 2.9 m and only change insubstantially within a 1.3–2.9 m propagation range, figure 3.11(a).

Conclusions

Pulse compression down to about 50 fs from the initial 150 fs pulse is achieved. A robust soliton-like optical pulse having smooth propagation dynamics is obtained within the (3+1)D nonlinear envelope equation at the ionization-free regime by a proper choice of the pulse and the medium parameters. The evolution of the pulse parameters and shape is studied along the propagation distance until the pulse splitting begins appearing. At a given stage of the propagation, the pulse shows stability with respect to the pulse diameter, peak intensity, time duration and spatiotemporal shape. The pulse satisfies the criterion of stability of spatial soliton but still not the criterion of stability of temporal soliton. Generation of such a robust pulse is observed for the first time in the ionization-free regime at realistic physical conditions [28].

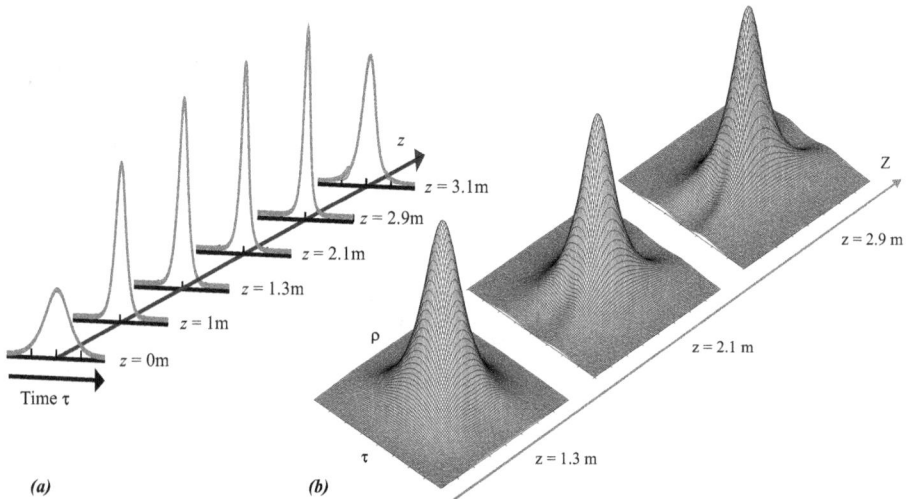

Figure 3.11. Pulse shape versus propagation distance. Time domain pulse shape (a), and spatiotemporal pulse shape (b) at ionization-free regime. Stabilization of the pulse profile between 1.3 m and 2.9 m takes place.

Ionization regime

Within the self-consistent model, equations (3.23)–(3.25), pulses as short as a single-cycle period, 2.5 fs, can be described at a strong regime of ionization. The 11-photon absorption is ruled by $\chi^{(21)}$-nonlinearity. Consequently, the leading nonlinearity in the (3+1)D NEE is $\chi^{(21)}$. The laser pulse energy is 0.5 mJ and the initial pulse shape is spatiotemporal Gaussian with a durationof 150 fs. The propagation medium is a pressurized argon at 18 atm pressure. The evolution of the main pulse parameters, beam radius, peak intensity, and the pulse duration are shown in figures 3.12–3.14, respectively. The beam diameter and the pulse duration are characterized as a FWHM of the pulse. The pulse rearrangement initiates with self-focusing, figure 3.12. It leads to a radial flow of energy towards the center of the pulse, which results in a growth of the peak intensity, figure 3.13. The latter, together with the negligible longitudinal expansion of the pulse over the propagation distance under consideration, lead to an effective shortening of the pulse, figure 3.14, due to the mechanism specified in [11]. In this relatively short initial stage (that covers typical laboratory scale distances of about one meter or less) and due to the relatively long initial pulse used here (150 fs pulses in comparison to 20–30 fs pulses used in more recent works [1, 9]), the pulse develops mainly within the minimal set of lowest order processes, as in the NLSE [11], and the pulse does not yet show signature of stabilization. The higher order terms in the NEE, equation (3.23), have little influence on that stage. With the increase of the peak intensity, the contribution of the quintic ($\chi^{(5)}$) nonlinearity of neutrals rises rapidly, and, due to the opposite sign to the cubic ($\chi^{(3)}$) one, starts saturating the nonlinearity. In reality, all orders of nonlinearity are involved in the saturation process [15]. The substantial changes of the pulse propagation dynamics initiate with increasing of the peak intensity around and above the level of substantial ionization of the medium. It triggers another set of processes in the NEE—the ionization terms. The beam diameter also shows

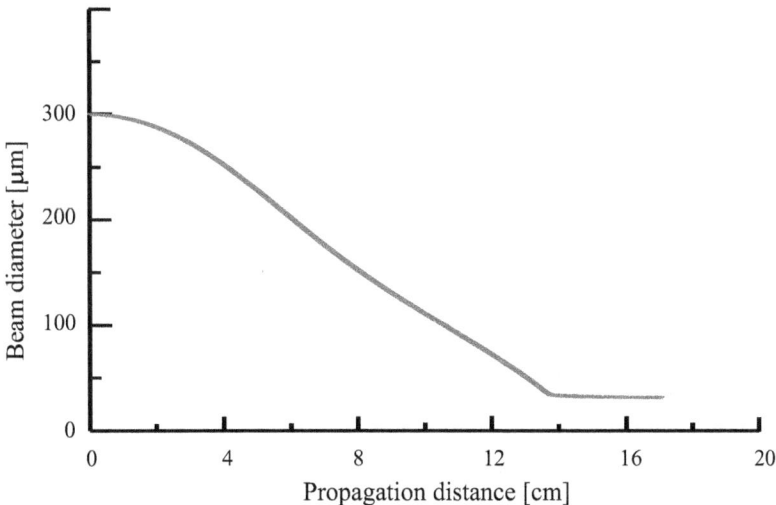

Figure 3.12. Beam diameter versus propagation distance.

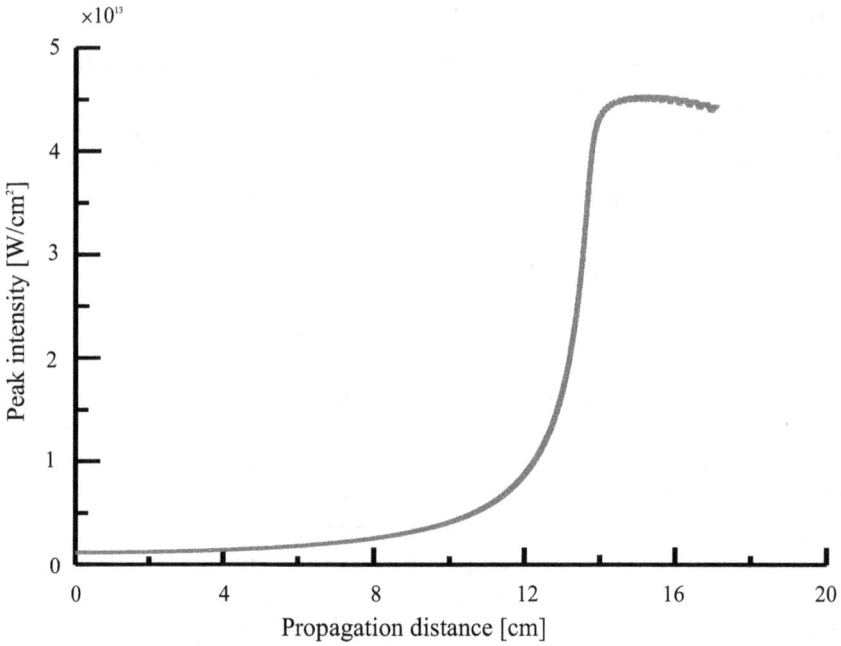

Figure 3.13. Peak intensity versus propagation distance.

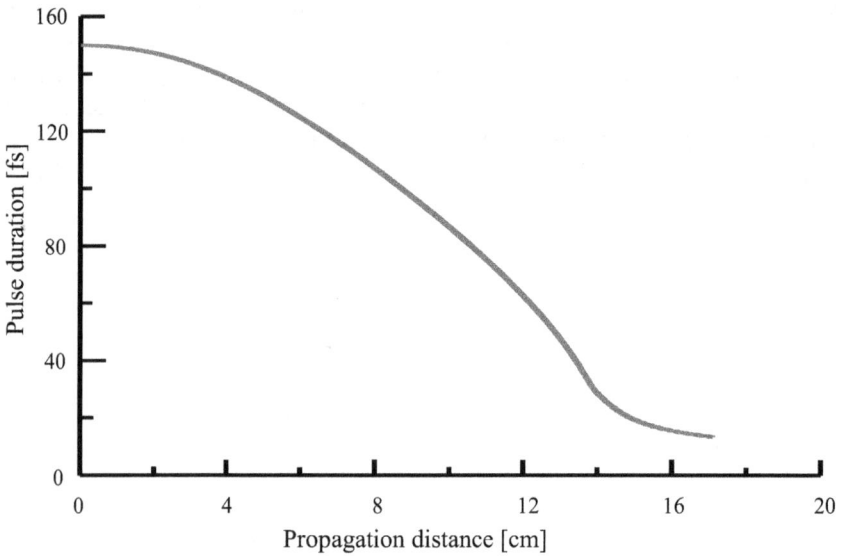

Figure 3.14. Pulse duration versus propagation distance.

stabilization along the same propagation range, figure 3.12. It results from a balance of self-focusing due to $\chi^{(3)}$-nonlinearity (and higher orders of self-focusing nonlinear terms), from one side, and defocusing due to the common action of diffraction (the most universal defocusing effect that, however, has a relatively weak contribution at the usual conditions of filamentation) of high-intensity ultrashort light pulses, $\chi^{(5)}$-nonlinearity (and higher orders of defocusing nonlinearities), and ionization, from the other. Such a balance stays behind the formation of a stable light filament (and spatial solitons, as well). Recently, filamentation without ionization has been proposed as a result of a dynamic balance between the focusing ($\chi^{(3)}$, $\chi^{(7)}$, …) and defocusing ($\chi^{(5)}$, $\chi^{(9)}$, …) nonlinearities [1]. Although the possibility of formation of stable light filament without ionization is well illustrated, the whole phenomenology that accompanies the filamentation, including the more important (in our opinion) effects, i.e. the SC of the pulse and, eventually, its stable propagation, are not demonstrated within such a mechanism. The contribution of higher orders of nonlinearities, together with the ionization, as a possible mechanism of the self-compression and spatio-temporal stabilization of the pulses, has been proposed in [12]. The evolution of the pulse in the time domain is illustrated by the time duration, figure 3.14, and on-axis pulse shape, figure 3.15. In the area of stabilization of the beam radius and peak intensity, the pulse also shows a signature of stabilization in time, although some slow systematic decrease of the pulse duration still exists. In that respect, the stabilization of the on-axis temporal pulse shape is better expressed, figure 3.15—see the time profiles of the pulse taken at the beginning ($z = 13.9$ cm), in the middle ($z = 14.5$ cm), and at the end ($z = 15.2$ cm) of the range of stabilization with respect to time. As can be expected, the range of stabilization of the pulse in time is most short in comparison to the other two ranges (in the transversal

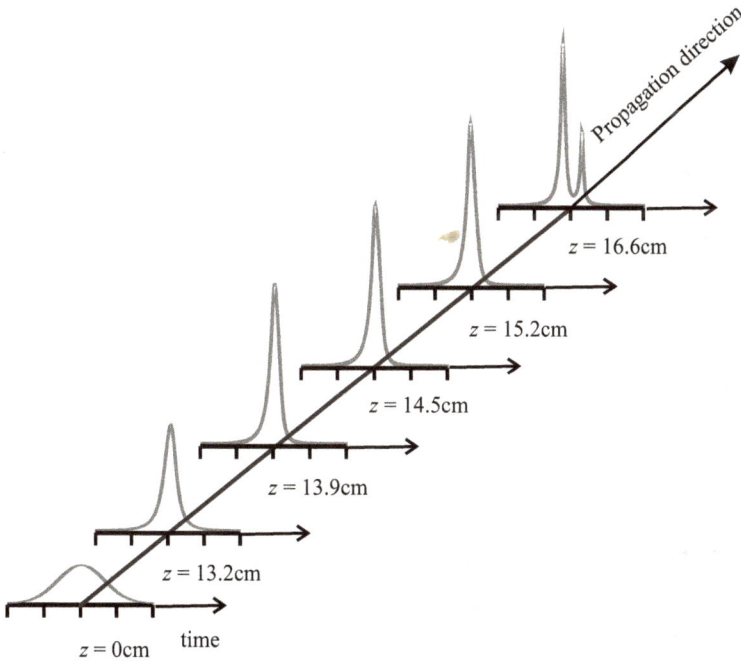

Figure 3.15. On axis profile of the pulse at strong ionization regime. Stabilization of the pulse profile between 13.9 cm and 15.2 cm takes place.

direction, figure 3.12, and in the peak intensity, figure 3.13) because it depends strongly on the stabilization of the pulse with respect to the other two parameters. The self-compression of the pulse from 150 fs (the initial pulse) to about 18 fs (at the end of the range of stabilization in time) takes place. At longer distances, a second pulse starts appearing at the trailing edge of the main pulse in agreement with the experimental observations—see the time profile at $z = 16.6$ cm in figure 3.15. Due to such deterioration of the pulse, the results in figures 3.12–3.15 are cut slightly after the region where it takes place. At the same time, the transversal profile of the pulse keeps its smooth bell-shape structure unchanged during the propagation. Based on the above result, we may conclude that a general stabilization of the pulse takes place over a propagation distance of length between 1.5 cm and 3 cm for different pulse parameters. Such an understanding is additionally enforced if we compare the behavior of the pulse predicted by the NLSE [11, 24]. In that case, the pulse immediately starts splitting after the peak intensity reaches maximum and none of the pulse parameters show stabilization. It means that the above-cubic nonlinearities and ionization play a substantial role not only in the pulse compression but also in the stabilization of the pulse parameters.

Pulse propagation dynamics without dispersion contribution

To reveal the role of dispersion in the pulse propagation, all dispersion terms have been neglected in the model [27]

$$\frac{\partial A}{\partial z} = \frac{i}{2k_0} \hat{T}^{-1} \nabla_\perp^2 A + i\frac{\omega_0}{c} n_2 \hat{T} |A|^2 A - i\frac{\omega_0}{c} n_4 \hat{T} |A|^4 A$$

$$- i\frac{k_0}{2n_0^2 \rho_c} \hat{T}^{-1} \rho A - \frac{\sigma}{2} \rho A - \frac{\beta_{\mathrm{MPI}}(A)}{2} A.$$

It leads to an insubstantial change of the pulse behavior. Consequently, the propagation dynamics of the pulse at the conditions under consideration is ruled mainly by the interplay of nonlinear processes of different order in neutrals and plasma.

Conclusion

Pulse compression down to about 20 fs from the initial 150 fs pulse is achieved. Spatiotemporal stabilization of the pulse over a given propagation distance is achieved at the presence of strong nonlinear processes, including ionization of the medium.

3.5 Summary of the studies

- The spatiotemporal dynamics of high-intensity ultrashort laser pulses is studied in three basic propagation regimes at realistic physical conditions based on the (3+1)D nonlinear Schrödinger equation and the (3+1)D non-linear envelope equation.
- Self-compression of the pulse is observed in all three cases.

- No stabilization of the pulse has been obtained for the case of the (3+1)D nonlinear Schrödinger equation.
- Pulse stabilization has been obtained for the case of (3+1)D nonlinear envelope equation at both, ionization-free and ionization regimes.
- For the case of the (3+1)D nonlinear envelope equation, the following general conclusion can be made. The ionization leads to a shorter distance of pulse stabilization but the compression rate is much stronger in comparison to the ionization-free regime.

References

[1] Béjot P, Kasparian J, Henin S, Loriot V, Vieillard T, Hertz E, Faucher O, Lavorel B and Wolf J-P 2010 Higher-order Kerr terms allow ionization-free filamentation in gases *Phys. Rev. Lett.* **104** 103903

[2] Berge L, Skupin S, Nuter R, Casparian J and Wolf J-P 2007 Ultrashort filaments of light in weakly-ionized, optically-transparent media *Rep. Prog. Phys.* **70** 1633–713

[3] Brabec T and Krausz F 1997 Nonlinear optical pulse propagation in the single-cycle regime *Phys. Rev. Lett.* **78** 3282–5

[4] Brabec T and Krausz F 2000 Intense few-cycle laser fields: frontiers of nonlinear optics *Rev. Mod. Phys.* **72** 545–91

[5] Chernev P and Petrov V 1992 Self-focusing of light pulses in the presence of normal group-velocity dispersion *Opt. Lett.* **17** 172–74

[6] Christov C I 1994 Gaussian elimination with pivoting for multidiagonal systems *Internal Report* 4 University of Reading

[7] Christov C I, Dost S and Maugin G 1994 Inelasticity of soliton collisions in systems of coupled NLS equations *Phys. Scr.* **50** 449–54

[8] Diddans S A, Eaton H K, Zozulya A A and Clement T S 1998 Amplitude and phase measurements of femtosecond pulse splitting in nonlinear dispersive media *Opt. Lett.* **23** 379–81

[9] Gaarde M B and Couairon A 2009 Intensity spikes in laser filamentation: diagnostics and application *Phys. Rev. Lett.* **103** 043901

[10] Koprinkov I G 2004 Ionization variation of the group velocity dispersion by high-intensity optical pulses *Appl. Phys.* B **79** 359–61

[11] Koprinkov I G, Todorov M D, Todorova M E and Todorov T P 2007 Self-compression of high-intensity femtosecond laser pulses in low dispersion regime *J. Phys. B: At. Mol. Opt. Phys.* **40** F231–6

[12] Koprinkov I G, Suda A, Wang P and Midorikawa K 2000 Self-compression of high-intensity femtosecond optical pulses and spatiotemporal soliton generation *Phys. Rev. Lett.* **84** 3847–50

[13] Marchuk G I 1986 Mathematical models in environmental problems *Studies in Mathematics and its Applications* ed J L Lions *et al* vol 16 (Amsterdam: North-Holland) pp 82–9

[14] Mlejnek M, Wright E M and Moloney J V 1998 Femtosecond pulse propagation in argon: a pressure dependence study *Phys. Rev.* E **58** 4903

[15] Nurhuda M, Suda A and Midorikawa K 2008 Generalization of the Kerr effect for high intensity, ultrashort laser pulses *New J. Phys.* **10** 053006

[16] Ranka J K and Gaeta A L 1998 Breakdown of the slowly varying envelope approximation in the self-focusing of ultrashort pulses *Opt. Lett.* **23** 534–6

[17] Ranka J K, Schirmer R W and Gaeta A L 1996 Observation of pulse splitting in nonlinear dispersive media *Phys. Rev. Lett.* **77** 3783–6

[18] Rothenberg J E 1992 Pulse splitting during self-focusing in normally dispersive media *Opt. Lett.* **17** 583–95

[19] Silberberg Y 1990 Collapse of optical pulses *Opt. Lett.* **15** 1282–4

[20] Tanev S and Pushkarov D 1997 Solitary wave propagation and bistability in the normal dispersion region of highly nonlinear optical fibers and waveguides *Opt. Commun.* **141** 322–8

[21] Todorov M D 2016 The effect of the elliptic polarization on the quasi-particle dynamics of linearly coupled systems of nonlinear Schrödinger equations *Math. Comput. Simul* **127** 273–86

[22] Todorov M D 2016 On a method for solving of multidimensional equations of mathematical physics *CSNDD Int. Conf. Structural Nonlinear Dynamics and Diagnosis* vol 83 Marrakech, Morocco ed M Belhaq (France: MATEC Web of Conferences, EDP Sciences) 05012

[23] Todorov M D and Christov C I 2007 Conservative numerical scheme in complex arithmetic for coupled nonlinear Schrödinger equations *Discrete Cont. Dyn. Syst. Suppl.* 982–92

[24] Todorov M D, Koprinkov I G, Todorova M E and Todorov T P 2008 Self-compression and controllable guidance of multi-millijoule femtosecond laser pulses *Opt. Commun.* **281** 5249–56

[25] Todorov T P, Todorova M E, Todorov M D and Koprinkov I G 2010 Generation of stable (3+1)-dimensional high-intensity ultrashort light pulses AMiTaNS'10 AIP CP1301 eds M D Todorov and C I Christov (Melville, NY: American Institute of Physics) pp 587–94

[26] Todorov T P, Todorova M E, Todorov M D and Koprinkov I G 2014 On the stable propagation of high-intensity ultrashort laser pulses *Opt. Commun.* **323** 128–33

[27] Todorova M E, Todorov T P, Todorov M D and Koprinkov I G 2014 Spatiotemporal dynamics of high-intensity ultrashort laser pulses in strongly nonlinar regime *22nd Int. Conf. Nonlinear Dynamics of Electronic Systems* vol 438 *Communications in Computer and Information Science* eds V M Mladenov and P C Petrov (Switzerland: Springer) pp 165–72

[28] Todorova M E, Todorov M D, Todorov T P and Koprinkov I G 2015 Robust ultrashort laser pulse formation in ionization-free regime propagation *AMiTaNS'15* AIP CP1684 ed M D Todorov (Melville, NY: American Institute of Physics) 080012

[29] Yanenko N N 1971 *Method of Fractional Steps* (New York: Gordon and Breach) pp 148–61

[30] Zozulya A, Diddams S, Van Engen A G and Clement T S 1999 Propagation dynamics of intense femtosecond pulses: multiple splittings, coalescence, and continuum generation *Phys. Rev. Lett.* **82** 1430

www.ingramcontent.com/pod-product-compliance
Lightning Source LLC
Chambersburg PA
CBHW061816210326
41599CB00034B/7017